FM 25-7
Pack Transportation
U.S. Army 1944

Printed in the United States of America

First Printing, 2020
ISBN 978-1-7353222-2-3

Carentan Media Group
2215 Plank Rd Suite 165
Fredericksburg, VA. 22401

Introduction

For centuries civilizations and armies utilized pack transport to move goods and people from point A to point B. With the advent of the gasoline reciprocating engine, the need for this was dramatically reduced…but never eliminated. Despite our best efforts, despite road building and aviation assets, there are still places where it will be cheaper and easier to just pack it in on mules and horses. This has proven itself time and again, from the forest fires of the west to the mountains of Afghanistan.

This was recognized in 2004 with the Army publication of FM 3-05.213 and again in 2014 with the revisions in Army Training Publication (ATP) 3-18.13, Special Forces Use of Pack Animals.

With the USFS out west, they never entirely did away with their pack trains. They maintain one in Lolo Peak National Forest in MT at the 9 Mile Ranger Station and another in the Sierras in California. They have even established a Pack Stock Center for Excellence in USFS Region 5 to bring up the next generation of packers.

WW2 and Korea was the last heyday of the military pack trains. The army had maintained institutional knowledge and remount centers up through the years, only closing the last pack unit at Ft. Carson in 1956. It would take almost fifty years for pack stock to come back on the Army's radar.

The last years, however, were an interesting time. Despite mass mechanization in WW2 they found that they couldn't entirely do away with the animals. In North Africa, Italy and Burma their contributions were essential.

"Pack mules were used by US forces in Tunisia during the winter of 1942-43 and were employed extensively in the rugged mountain terrain of Italy. In the hard mountain fighting before Cassino, pack animals were extensively used. During 1944-45, the 10th Mountain Division employed over 14,000 mules in the rugged terrain of northern Italy during its drive through the North Apennines Mountains and the Po Valley.

"Animal pack outfits were also used in the China-Burma-India theater especially during the active combat operations in Burma. Very often the Army would procure animals in the theaters where the troops were operating and in emergencies would commandeer animals on the spot," writes Oliv-Drab.com.

"Unconventional forces in Burma (now Myanmar), including Merrill's Marauders, used mules quite effectively. During the operations against Myitkyina (pronounced Mish-i-naw), the key objective in Northern Burma, the 475th Infantry Regiment, 124th Cavalry, two battalions of pack artillery and QM Pack Troops became the Mars Task Force.

A Liberty ship brought about 275 mules of the 35th Pack Troop to India, then to Ledo by train. An overland march 300 miles down the Ledo Road brought them to Camp Landis, Burma. The mules were divided among the units of the force and served to carry machine guns, mortars, ammunition and other supplies in terrain where no other method was feasible.

On 17 May 1944, Merrill's Marauders took their pack mules on a 65 mile march traversing the Kumon Mountain range to attack and capture the airfield at Myitkyina, the only all-weather airfield in the country, and the gateway to the road to China."

In Korea it was much the same; terrain where access to wheeled vehicles was near impossible in places. While the U.S. Army had no formal horse or mule detachments at this time, they were fortunate enough to capture large numbers of Chinese stock during the U.N. counterattack north from Seoul in late May 1951. These animals were quickly snapped up by the 1st Cavalry division who established a remount depot to the south. The animals were rehabilitated and returned to service by the Army.

One captured mule of note was found to have a standard U.S. Army Quartermaster brand; Preston Brand number 80K0. Believed to have been part of the Mars Task Force he was thought to have been captured from Chinese Nationalists forces by the communists, as the Mars unit had turned their mules over at the end of the war. His long journey eventually returned him to US custody and he spent the rest of the conflict carrying things for the GIs again.

With the Korean War wrapping up, however, the age of the Army pack trains came to an end. The helicopter superseded them and there was no use for them, thankfully, in the highlands of Vietnam. That would have to wait until America's post 9-11 intervention in Afghanistan.

This Manual

The manual contained herein was one of the last official Army publications to be printed that dealt with managing pack trains and their animals. Only with the SF version in 2004 and 2014 did the Army return to the subject, and then only in a Special Forces context.

It represents the cumulative knowledge of almost two hundred years of mule and horse packing in the US military.

While the only pack rig it details is the three types of standard issue Phillips pack saddles, the lashing, knot work, animal maintenance and management of mass numbers of animals is invaluable.

Also of interest is the methodology for fashioning pack saddles into boats for water crossings. The mules, of course, simply swam.

* A note on the photos; the original scan of the surviving document was done on somewhat primitive equipment. It washed them out badly. The line drawings, however, are pristine.

≈. D. HASTINGS

FM 25-7

WAR DEPARTMENT FIELD MANUAL

PACK

TRANSPORTATION

WAR DEPARTMENT ● 25 AUGUST 1944

WAR DEPARTMENT FIELD MANUAL
FM 25-7

This manual supersedes Chapter 5, FM 25-5, Animal Transport, 15 June 1939, and Chapter 4, FM 6-110, Pack Artillery, 1 March 1940.

PACK
TRANSPORTATION

WAR DEPARTMENT ● *25 AUGUST 1944*

United States Government Printing Office
Washington : 1944

WAR DEPARTMENT,

WASHINGTON 25, D.C., 25 AUGUST 1944.

FM 25–7, Pack Transportation, is published for the information and guidance of all concerned.

[A. G. 300.7 (19 Jul 44).]

BY ORDER OF THE SECRETARY OF WAR:

G. C. MARSHALL,
Chief of Staff.

OFFICIAL:

J. A. ULIO,
Major General,
The Adjutant General.

DISTRIBUTION:

Base Comds (2); All Sv C (2); Island Comds (2); Def Comds (2); Depts (2); Harbor Def (2); Armies (2); Corps (2); D 2, 6, 7, 10 (2); B 2, 6, 7, 10 (2) except I B 6 (6); R 2, 6, 7, 10 (2); Bn 2, 6, 7, 10 (2) except I Bn 6 (6); I C 2, 6, 10 (15); Remount Deps (50).
I B 6:   T/O & E 6–270 T;
I Bn 6: T/O & E 6–155;
I C 2:   T/O & E 2–17; 2–19;
I C 6:   T/O & E 6–156; 6–157;
I C 10: T/O & E 10–97; 10–118.

For explanation of symbols, see FM 21–6.

CONTENTS

This manual supersedes Chapter 5, FM 25-5, Animal Transport, 15 June 1939, and Chapter 4, FM 6-110, Pack Artillery, 1 March 1940.

CHAPTER 1
GENERAL

1. MISSION. The mission of pack transportation is to transport loads on the backs of animals over terrain which is difficult for or impassable to wheeled or track-laying vehicles. Its success depends largely upon the careful selection and training of personnel and pack animals. The employment of correct packing and march techniques is essential.

2. CLASSIFICATION.

a. Pack transportation facilities are of three distinct types:

(1) Cargo pack trains operated by quartermaster corps, pack artillery, infantry, and engineer units using organic means. Loads, generally bulky and heavy, are secured to saddles by ropes. Gaits are the walk and amble.

(2) Artillery combat pack units, using organic means. Loads, such as heavy howitzer parts, instruments, communication equipment, and ammunition, as well as regular cargo loads, are secured to the saddle with arches, adapters, and hangers. Gaits are the walk and amble.

(3) Horse cavalry, using organic means. Loads of reduced weight and bulk are packed in hangers and carriers and so positioned as to enable the animals to maintain equilibrium at the walk, trot, and gallop. Special cargo loads are secured to the saddle with ropes.

b. Cargo pack trains and artillery combat pack units use mules. Cavalry may use either mules or horses.

For military terms not defined in this manual, see TM 20–205.

c. It is essential that pack transportation facilities be so maintained as to be capable of continuous operation. For this work, skilled personnel is required.

3. CHARACTERISTICS AND CAPABILITIES.

a. Pack transportation provides a reasonably rapid, quiet, and reliable mobility in mountains, jungles, and other terrain unsuitable for vehicular transportation.

b. Pack transportation units are not organized, trained, or equipped to operate on roads, highways, deserts, or in deep snow. The physical condition of animals is materially impaired by long rail, truck, or boat trips; consequently, the need for pack transportation should be anticipated sufficiently in advance to permit proper conditioning of the animals prior to their employment in campaign.

c. Over terrain which is not mountainous, the pack mule may be expected to travel 20 miles or more per day carrying 250 pounds of pay load. (Pay load does not include weight of the saddle and its accessories.) As long as the mule receives proper care and feed, this expectancy of his capability continues indefinitely. In mountainous terrain, the mule is capable of carrying 250 pounds, but the distance should be reduced to 10 or 15 miles per day. Loaded pack mules usually are able to travel anywhere a man can walk without the use of his hands for support.

4. TRAINING OF PACK TRANSPORTATION UNITS.

a. It is imperative that pack transportation units be trained on the type of terrain over which they are to operate.

b. Mobility of pack transportation depends largely on three factors:

(1) Selection and training of quiet, gentle, and manageable animals.

(2) Ability of personnel to care for and pack the animals so as to obtain the maximum use of them.

2

(3) Physical condition of both men an animals. Training, therefore, should include a carefully planned and executed remount training program, extensive practice in packing all types of loads, and marching of the animals under full load over all types of terrain. Conditioning can be acquired only by daily marching of men and animals over varied terrain. Good march discipline, a thorough knowledge of pack transportation, and careful supervision of the march are essential to success.

CHAPTER 2
SELECTION AND TRAINING OF PACK ANIMALS

---

5. SELECTION.

a. For artillery and quartermaster pack transportation, pack mules are issued as such. Cavalry must select its pack animals from those issued for riding.

b. In general, a pack mule should be from 14¾ to 15½ hands in height and weigh from 1,000 to 1,200 pounds. He should be compact, stockily built, and have a short neck; short, straight, strong, and well-muscled back and loins; low withers and croup; large barrel with deep girth; straight, strong legs; and short pasterns and good feet.

c. In addition to desirable physical proportions, pack animals should be gentle and have friendly dispositions. They should have no fear of man and should be free of vices and vicious habits. They should walk and trot freely and boldly over varied terrain. There should be little movement of the back and a minimum of side swaying of the body while the animal is in motion.

d. The defects of conformation to be avoided in the selection of pack animals are:

(1) Withers—too thick, too flat, or too thin.
(2) Back—too short, too long, swayed, or roached.
(3) Chest—broad-ribbed, draft type.
(4) Barrel—excessively large.

e. Horses for use under pack are selected from all the horses of the organization. If practicable, all pack horses should have completed basic remount training as described in FM 25-5. When pack horses have been selected and drivers assigned (if the pack horses are to be driven or led by troopers), the horses for drivers are selected. It is desirable to pair pack and riding horses so as to insure a smooth-working team.

6. TRAINING.

a. General. In the training of new pack animals, the principles set forth in FM 25–5 apply, particularly those which pertain to the conditioning of animals and to the use of quiet, patient, and persistent methods of instruction. Any system of training that neglects the conditioning or destroys the tranquility of the new pack animal is defective. *The mobility of pack transportation in the field depends in a great measure upon the gentleness and willingness of the pack animals.*

b. Selection of trainers. Proper selection of personnel to train new pack animals is extremely important. The men should be selected because of their knowledge and lack of fear of animals. Their personal qualities should include patience, kindliness, and firmness.

c. Gentling.

(1) Fear is one of the animal's strongest instincts. If it is allowed to remain the dominant instinct, the animal cannot be trained satisfactorily to do the work demanded of him. Throughout the training period, the goal of all concerned should be to gain the confidence of the animal.

(2) Rewards for accomplishment are extremely valuable in the gentling process. Patting the neck, rubbing the head, and hand-feeding are good aids in gaining the confidence of the animal. The use of whips, twitches, or uncontrolled enthusiasm should not be allowed in the training.

d. Leading. All pack animals must be taught to lead. One method is to lead them alongside well-broken animals. Leading should be at the walk as a daily exercise until new animals lead quietly and·have improved

5

sufficiently in condition to allow them to undergo instruction under the saddle. If at first the animal does not lead readily, the use of a haunch rope, a hand-offering of grain, or a combination of those expedients, will prove effective in a majority of cases.

e. Riding. All mules should be broken to riding and ridden regularly during training prior to work under the pack. Since most mules' mouths are tender, bits should not be used during the initial riding periods. Reins may be attached to the halter or the hackamore may be used.

f. Standing. Mules should be taught to stand quietly when the rider dismounts and drops his reins to the ground. This can be accomplished as follows: the rider attaches the end of a lair rope to the halter and coils the remainder of the rope on the saddle horn; to bring the mule to a halt, the rider calls "Whoa," drops the split reins to the ground, dismounts quickly and, carrying the coiled lair rope in the hand, moves quietly away from the animal, paying out the rope. He arrests any movement of the animal by a quick tug on the lair rope. This process should be repeated until the mule will stand when the reins are dropped. The lair rope aid is then removed. For further training, the rider, upon dismounting, ties the halter shank to a stake or other object on the ground. This will discourage movement of the animal and cause him to stand even when the rider moves out of his sight.

g. Packing.

(1) After having been ridden for about 10 days, the animal should be saddled with the pack saddle. For the first few days, the pack saddle should not be loaded. Thereafter, the animal should be packed with a single load, such as a sack of oats, and the load increased progressively until he is carrying a full pay load of approximately 200 to 250 pounds.

(2) All pack animals should be taught to stand quietly while being saddled and packed. If at first an animal

6

will not stand quietly, the blind should be put in position. The blind is an exceptional aid and should be used only when the need is clearly indicated. *The animal must never be moved a single step while blinded.*

h. Bell mare. Mules, being hybrid animals, show a definite fondness or affection for a mare and, to a lesser degree, a gelding. A bell mare should be kept with the mules in training at all times. This practice tends to make the mules more docile and easier to handle. A bell is attached to the mare with a neck strap and worn constantly. The mules will associate the sound of the bell with the presence of the mare.

i. Herding.

(1) All pack mules must be taught to move quietly in a herd following the bell mare. She always is led at the head of the column by a mounted man. Training in herding may be begun as part of the exercise of the pack animal. Any available or improvised oval-shaped track may be used. The bell mare is led around the track, and the pack animals are kept on the move behind her by several drivers. This exercise should be accomplished quietly and, at first, slowly, to avoid exciting the animals. *There should be no shouting and cracking of whips.* The herd then should be conducted out of the track area as follows:

(*a*) Two or more packers riding at the head of the column to prevent the herd from passing the bell.

(*b*) Several packers riding each side of the column to prevent the animals from wandering and to adjust loads when necessary.

(*c*) One packer riding at the rear of the column to prevent straggling and undue extension of the column.

(2) During such practice marches, advantage should be taken of narrow trails and defiles to teach the animals to march in single file without crowding.

(3) Because pack horses do not herd as well as mules, they seldom are herded.

j. Training for proper gaits.

(1) The best gaits for pack mules are the walk (approximately 4 miles per hour) and the amble (5 miles or slightly more per hour). Since the trot and gallop usually derange the loads and fatigue pack mules, such gaits seldom are used. Under exceptional circumstances, the trot may be employed for short distances. With average loads, the gallop is taken only in an emergency and for very short distances only. During training periods, the pace of the walk should be extended gradually. Herded pack mules, in their endeavor to keep up with the bell mare, frequently will break into the amble for a few steps and gradually become confirmed in this gait.

(2) Pack horses are trained to carry their loads at all of the gaits used in marching and maneuver. The walk and trot are used habitually, but the gallop is used only when the tactical situation makes it expedient to do so.

k. Swimming.

Pack animals must be taught to swim boldly and freely. Although they are naturally good swimmers, some animals are afraid of water and will resist entering it. When in the water, such animals fight it and swim very poorly. All animals should be introduced to water quietly, coaxed to wade through shallow water at first, the depth being gradually increased until they must swim. The presence of known good swimmers during this training will lend confidence to the green swimmer.

l. Battle inoculation.

During training, pack animals should be mentally conditioned to as many as possible of the sights, noises, and odors common to combat zones. Once the animals become familiar with these sensations, their docility and good conduct in the field will be assured. This mental conditioning, or *battle inoculation*, must be so conducted that animals will not associate the sights, noises, and odors with harm or pain to themselves. A few specific suggestions are as follows:

8

(1) Animals should be conducted to the vicinity of motor parks while engines of vehicles therein are being warmed up. They should be familiarized with the sounds of vehicle exhausts, metal being pounded against metal, and with the sight of tractors, tanks, and heavy vehicles.

(2) Pack animals should become familiar with sounds of weapons firing, locomotive whistles, and low-flying aircraft.

(3) Cans containing pebbles may be attached to loads during the training period. Pack animals must be taught not to fear creaks and rattles. Many of the loads which they subsequently will be required to carry will produce odd noises.

(4) Pack animals should be subjected to a variety of odors such as iodine, ether, smoke, gasoline, disintegrat· ing flesh, and rotting vegetation.

(5) To strengthen the pack animal's sense of balance (equilibrium), he should be made to cross extremely narrow bridges, fallen trees and ditches. He should be worked on steep narrow trails and corduroy roads over swamps and bogs. Narrow bridges may be simulated by constructing a gang-plank walk, approximately 2 feet wide and initially not to exceed 1 foot in height, over a pool of water or small stream bed. Subsequently, narrower and higher bridges should be constructed.

(6) Pack animals must be able to walk up and down short inclines and, when necessary, take slides confidently and without hesitation.

(7) Top-load animals should be conditioned under full pay loads before being required to carry howitzer loads.

7. SUGGESTED 21-DAY PACK MULE REMOUNT TRAINING SCHEDULE.

a. General. The gentling, training, and conditioning of remount pack mules should be begun as soon as possible following their receipt from a remount depot or other source. They are placed immediately in quarantine for a period of not less than 21 days. This period may be used advantageously for the initial gentling and training of the animals, provided their state of health will permit. Newly-arrived remounts usually are nervous, soft, fat, underworked and, possibly, overfed. To fit them for field service they must be gentled, hardened, seasoned, and accustomed to hard work. Early phases of this training should be conducted for remounts as a group.

b. Objective. The primary objective of remount training is to produce gentle and willing animals. Abuse; whips, blinds, and rough treatment will defeat the purpose. The use of a snub animal is permissible when necessary.

c. Detailed daily schedule.

FIRST DAY

First hour. Check Preston brands against the mule record cards. Check equipment to be used and place it in quarantine for the 21-day period. Instruct handlers to tie animals to well-seated posts or other sturdy objects, and not to the boards of a fence. *It is essential that animals not be permitted to break away during training.*

Second hour. Select personnel to work with the remounts. Assign each man definitely to a particular animal. Nervous animals should be assigned to the best available trainers.

Third hour. Inspect each animal for signs of shipping diseases. Instruct handlers to report animals which have nasal discharge, watery discharge from the eyes, signs of ringworm, or other symptoms. Place suspected animals on the sick book for disposition by the veterinary officer.

Fourth hour. This is the "get acquainted" hour. Learn everything possible about the men and animals. To win the confidence of their animals, handlers should talk to them, hand-feed and pet them. In doing so, trainers should be quiet, patient, and firm. Hand-feeding usually will teach the animal not to fear and cause him to associate his handler with feed. As a reward, hand-feeding is one of the best methods of gaining the animal's confidence.

Fifth hour. With a quiet animal, demonstrate the method of halter breaking with the aid of a haunch rope.

Sixth hour. Demonstrate halter breaking by use of the hackamore or a halter with a neck rope. Before attempting to use bits, employ one of the foregoing until the animal responds to the rein aids. A stubborn animal that will not lead should be snubbed to the saddle horn of the bell mare and pulled along. (The bell mare may be ridden in the corral for this purpose; she is ridden at no other time.) Have the handler walk in front of the animal with the halter shank attached to the halter and simulate leading the animal. A few days of such treatment usually will convince the animal that his handler is capable of leading him. On stubborn animals, or those that break away from their handlers, use of a neck rope with a half-hitch around the head just above the nostrils is an effective restraint and control measure.

Seventh hour. This is the initial grooming and gentling period. Work on the animal, using the brush lightly and hand rubbing. Limited use of the currycomb is appropriate. Clean out the front feet and simulate shoeing them by tapping lightly with a stick or small stone.

Eighth hour. Discuss the day's work, pointing out faults and appropriate corrective measures therefor; in some manner, by incident or example, bring out the importance of patience, kindness, and firmness in handling pack animals. Outline the schedule for the following day.

11

SECOND DAY

First hour. Lightly exercise the animals by walking them around the corral. Have a man ride the bell mare and have all other men form with ropes an inside fence similar to a race track. Have one man drive the animals from the rear. Stress quietness.

Second hour. Tie up all animals for health inspection by the veterinarian.

Third hour. Breaking and leading. Use a haunch rope if necessary.

Fourth hour. Introduction of animals to strange things such as ropes, saddle equipment, papers, raincoats, and blankets. Drag a rope around the animal's feet, between his legs, across his head and neck, and over his back. Let the rope fall over his haunches to the ground in the rear. *This must be done quietly.*

Fifth hour. Require trainers to mount animals bareback. Use the full hour. If the animals resist, use more time. Do not permit handlers to move the animals.

Sixth and seventh hours. Review the work of third, fourth, and fifth hours.

Eighth hour. As for first day.

THIRD DAY

First hour. As for second day.

Second hour. As for second day.

Third hour. Review training outlined for the third, fourth, and fifth hours of the second day.

Fourth hour. Introduction to the blanket and surcingle. Cinch lightly. Walk animals.

Fifth hour. Introduction to the packer's saddle. Use light front cinch pressure, remove the rear cinch, and leave stirrups hanging down. Lead animals.

Sixth hour. Tighten front cinch. Practice mounting and dismounting. Use a snub animal if necessary.

Seventh hour. Review work of third, fourth, and fifth hours. Include tying a raincoat loosely to the saddle, first on the pommel and then on the cantle. Repeat.

Eighth hour. As for first day.

FOURTH DAY

First hour. As for second day.

Second hour. As for second day.

Third hour. Review training outlined for the third, fourth, and fifth hours of the second day.

Fourth hour. Saddle and practice mounting and dismounting, near and off sides.

Fifth hour. Practice moving out at the walk and halting.

Sixth hour. Review training outlined for the third, fourth, and fifth hours of the second day.

Seventh hour. Mount and practice turns to the right and left.

Eighth hour. As for first day.

FIFTH DAY

First hour. Exercise animals at the walk and trot alternately.

Second hour. As for second day.

Third and fourth hours. Review previous work, especially phases in which individual animals appear to be deficient.

Fifth hour. Mount the animal and move out at the walk, letting a rope drag from saddle horn. Condition the animal to the feel of ropes by placing them on various parts of his body.

Sixth hour. Practice moving out at the walk, turning right and left, and halting.

Seventh hour. Review all previous training.

Eighth hour. As for first day.

SIXTH DAY

First hour. As for fifth day.

Second hour. As for second day.

Third, fourth, and fifth hours. Review all previous training.

Sixth hour. Require animals to drag loads of not to exceed 25 pounds. Make the animal pull straight away. Initially, handlers should be dismounted; later they should be required to mount and continue the exercise.

Seventh hour. Practice moving out, halting, and turning to the right and left at the walk only. While mounted, handlers should practice picking up objects from ground on both near and off sides.

Eighth hour. As for first day.

SEVENTH DAY

First hour. Grooming and health inspection.

NOTE: Remainder of day, rest. *If proper care has been used during this first week, no animals will have bucked or run away. The animals should be in good health and beginning to gain condition. Some of the animals received may have had a limited amount of training prior to purchase. Even though such animals show more response to the training, they should not be allowed to advance beyond the training schedule. Give a talk to all handlers on being gentle and firm. Commend handlers who did outstanding work during the week.*

EIGHTH DAY

First hour. Walk 30 minutes. Walk and trot alternately 30 minutes.

Second hour. As for second day.

Third and fourth hours. Review first week's training.

Fifth hour. Three-mile mounted march at the walk. Bell mare should be led at head of column.

Sixth hour. Using split reins and lair rope, teach animals to stand quietly.

Seventh hour. Mount and ride animals double. (Walk only.)

Eighth hour. As for first day.

NINTH DAY

First hour. As for eighth day.

Second hour. As for second day.

Third and fourth hours. Review appropriate phases of the first week's training.

Fifth hour. Three-mile cross-country ride, as individuals, at the walk and trot.

Sixth hour. As for eighth day.

Seventh hour. Lead from off side.

Eighth hour. As for first day.

TENTH DAY

First hour. As for eighth day.

Second hour. As for second day.

Third and fourth hours. Five-mile mounted cross-country march with bell mare at head of column.

Fifth and sixth hours. As for eighth day.

Seventh hour. As for sixth hour of eighth day.

Eighth hour. As for first day.

ELEVENTH DAY

First hour. Tie up animals for health inspection by veterinarian.

Second and third hours. Five-mile mounted cross-country ride as individuals.

Fourth hour. Review sixth hour of eighth day. Simulate saddling with pack saddle. Place riding saddles on animals from rear, two men working as a team.

Fifth hour. Require animals to drag a load of 100 pounds for 30 minutes. Ropes to be attached to saddle horns. During the remainder of the period, handlers should ride at will, carrying cans containing objects which rattle.

Sixth hour. Practice carrying loads on saddles and riding double, 30 minutes. Lead from off side, 30 minutes.

Seventh hour. Introduce animal to pack saddle. Saddle up and adjust cinches and breeching. Lead animals around the corral.

Eighth hour. Discuss the day's work, pointing out faults and appropriate corrective measures therefor; in some manner, by incident or example, bring out the importance of patience, kindness, and firmness in handling pack animals. Outline the schedule for the following day.

TWELFTH DAY

First hour. As for eleventh day.

Second and third hours. With animals under pack saddle, conduct a 3-mile march. Bell mare at head of column.

Fourth and fifth hours. Review appropriate phases of the first week's training.

Sixth and seventh hours. Review all of the second week's work.

Eighth hour. As for first day.

THIRTEENTH DAY

First hour. As for eleventh day.

Second and third hours. Saddle with pack saddle only. Make a 5-mile march with bell mare at head of column. Dismounted handlers lead alternately from near and off sides.

Fourth hour. Review sixth hour of the eighth day.

Fifth hour. Review seventh hour of tenth day.

Sixth hour. Review fifth hour of eleventh day.

Seventh hour. Lead animals over obstacles such as improvised bridges and ditches. Animals should not be allowed to jump. Hold their heads down.

Eighth hour. As for first day.

FOURTEENTH DAY

First hour. As for eleventh day.

NOTE: Remainder of day, rest. *Animals should be in good health and definitely on the way toward better condition. Goal: The ability to march 20 miles a day under full pay loads.*

FIFTEENTH DAY

First hour. As for eleventh day.

Second hour. Selection of animals for carrying top loads, side loads, and riding. Particular attention should be paid to the selection of animals for carrying top loads. Consider points of conformation such as straight back, low croup, and smooth gaits. Top-load animals should have plenty of bone. Mature animals are considered more desirable than others for this purpose.

Third and fourth hours. Three-mile march for pack animals under saddle. Riding mules to rest during this period.

Fifth, sixth, and seventh hours. Three-mile march for riding animals. Walk and trot alternately. Handlers to ride at will. Give animals additional training in standing quietly. Handlers to practice mounting from near and off sides. Make animals drag 100-pound weight. Ride double. Require packmasters and cargadors to start fitting saddles to pack animals.

Eighth hour. As for first day.

SIXTEENTH DAY

First hour. As for eleventh day.

Second, third, and fourth hours. Five-mile march over obstacles and slightly difficult terrain. Pack animals under saddle and a 50-pound single load. Riding animals rest.

Fifth, sixth, and seventh hours. As for fifteenth day (for riding animals). Require packmasters and cargadors to continue fitting pack saddles to pack animals.

Eighth hour. As for first day.

17

SEVENTEENTH DAY

First hour. As for eleventh day.

Second, third, and fourth hours. As for sixteenth day.

Fifth, sixth, and seventh hours. Five-mile march for riding animals only. Have packmasters and cargadors continue fitting saddles to pack animals.

Eighth hour. As for first day.

EIGHTEENTH DAY

First hour. As for eleventh day.

Second, third, and fourth hours. As for sixteenth day, except increase single load to 75 pounds.

Fifth, sixth, and seventh hours. For riding animals only, a 7-mile mounted march over obstacles and difficult terrain. Require packmasters and cargadors to continue fitting saddles to pack animals during this period.

Eighth hour. As for first day.

NINETEENTH DAY

First hour. As for eleventh day. Veterinarian to apply the mallein test.

Second, third, and fourth hours. For pack animals only, a 5-mile march under 100-pound single loads.

Fifth, sixth, and seventh hours. As for eighteenth day, except increase distance to 8 miles.

Eighth hour. As for first day.

TWENTIETH DAY

First hour. As for eleventh day.

Second, third, and fourth hours. As for nineteenth day, except increase distance to 8 miles.

Fifth, sixth, and seventh hours. As for eighteenth day, except increase distance to 10 miles.

Eighth hour. As for first day.

TWENTY-FIRST DAY

First hour. As for eleventh day. Veterinarian to read the mallein test.

Second hour. If no further quarantine is ordered, animals are released from quarantine and turned to duty with assigned units. Single out good handlers and commend them appropriately.

CHAPTER 3
PHILLIPS PACK SADDLES

Section I. TYPES OF SADDLES

8. GENERAL.

a. There are two *standard types* of the Phillips pack saddle.

(1) The *smaller type*, the cavalry pack saddle, is used for all cavalry and some infantry loads. It weighs 43 pounds (exclusive of breeching, cinches, and breast collar). Pads are 22 by 19 inches.

(2) The *larger type*, the cargo pack saddle, is used for all pack artillery loads, all pack train loads, and the heavier infantry weapon cargo loads. It weighs 72 pounds (exclusive of breeching and cinches). Pads are 23 by 25 inches.

b. A *third type*, the Phillips cavalry, modified (China special), is designed especially for smaller pack animals. The saddle, built to carry the same type of load as the larger cargo pack saddle, has the same external measurements as the cavalry pack saddle, but internally it is designed to fit an animal weighing approximately 800 pounds.

9. PHILLIPS CARGO PACK SADDLE. The saddle, consisting of a metal pack frame with detachable pads, is equipped with specially designed breeching, cinches, and woven pad. With these accessories, it weighs approximately 95 pounds.

a. Pads. There are two separate pads attached to each frame. Their primary function is to cushion the weight of the frame and load against the animal's back. The pads are attached to the frame by means of two

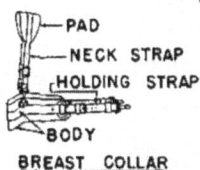

FIGURE 1. Phillips pack saddle, cavalry.

bronze lock staples near the top of each pad and by bottom bar pockets in the lower corners. The staples engage two lock staple hooks near the top of the inside of the frame. The bottom bar of the frame fits into the bottom bar pockets of the pads and is secured by bottom bar pocket pins. The outside or back of these pads is of leather reinforced internally with aluminum alloy ribs. There are five handholes in the back of each pad through which the padding may be quickly adjusted.' The bearing surface (part that comes in contact with the animal) of the pad is woven felt which has considerable stretching and gripping qualities. The pads are stuffed with curled hair which retains its resiliency indefinitely. This curled hair is kept in place within the pad by means of spaced leather thongs which, passing through the felt bearing surface, are tied on the outside of the leather surface. Thongs may be removed quickly when it is desired to adjust padding. To each lower corner of the pad is attached an aluminum bot-

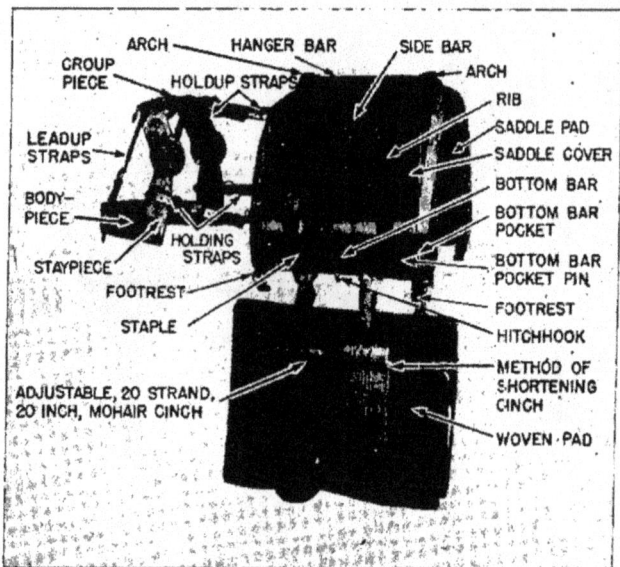

FIGURE 2. Phillips pack saddle, cargo, assembled.

tom bar pocket, an integral part of which is the footrest. These footrests protect the saddle pads by keeping them off the ground or floor when the saddle is not in use, and also they provide a hook over which to secure the quarter ropes of the hitches used in packing.

b. Frame. The metal frame serves two main purposes. It makes of the saddle a single unit, and at the same time provides an appropriate structure on which to pack all types of loads. The frame consists of steel arches, hanger bars, aluminum side bars, spring steel ribs, and aluminum bottom bars on which are attached three staples and one hitch hook.

(1) The *arches* are made especially strong to support heavy top loads. There are two holes in each arch proper for attaching the heavy type of arch or the adapter used for carrying top loads. A third hole, in the de-

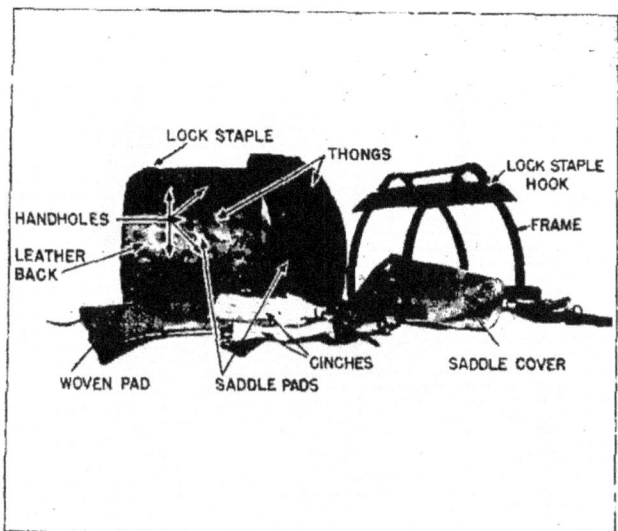

FIGURE 3. Phillips pack saddle, cargo, disassembled.

pression of the arch, facilitates attachment of the light type of load arch.

(2) The *hanger bars* which connect the arches give additional stability to the frame and provide a place on which to hook hanger loads.

(3) The *side bars* serve as a backbone for the entire frame and support most of the weight of the side loads.

(4) The *ribs* connect the side bars with the bottom bars and assist in supporting the weight of the side loads.

(5) The *bottom bar* aids in locking the frame to the pads and lends rigidity to the lower edges of the pad. The staples on the bottom bar are for strapping down hanger loads. The hitch hooks are used for securing the lash rope for the various type hitches. Ends of the frame are exactly the same, thus making it reversible.

c. Saddle cover. A leather bordered piece of canvas covers the pads of the saddle. The saddle cover protects

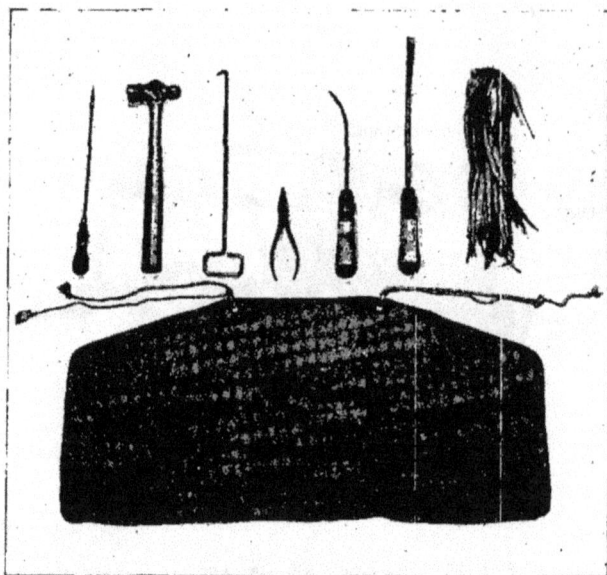

FIGURE 4. Tool roll, Phillips pack saddle.

the pads from rain and wear, keeps the heat of the sun from entering them, and shades the animal's spinal column.

d. Cinches. The saddle is equipped with two adjustable woven 20-strand cinches (fig. 2). The cinches (length 20 inches) may be shortened 2 inches by passing the small **D** through the bars of the large **D**, from outside to belly side, and smoothing out the folds. Further shortening is accomplished by placing a stick or piece of rope through the fold.

e. Cinching device. With each cinch, a cinching device is provided. This device is intended to make possible rapid cinching and release of cinches, accurate adjustment of cinch pressure, and to eliminate the tying and untying of knots. As a field expedient, these devices may be replaced by latigo straps.

f. Breeching. Each saddle is equipped with a breeching designed to prevent the saddle from riding forward on the animal's back. The breeching functions principally through two holding straps attached to the lower **D**-rings on the rear edge of the saddle. Separate body and croup pieces provide the main bearing surfaces. The stay piece connects the body piece to the croup piece and equalizes the pull exerted by the holding straps. The lead-up straps and hold-up straps serve primarily to hold the croup piece and the body piece in their proper positions. Metal buckle covers on these lead-up straps prevent the animal's tail from being caught in the tongues of the buckles.

g. Woven pad. A woven mohair pad, which readily shapes itself to the animal's back, serves as a cushion between the saddle pads and the animal's back, thus eliminating most of the friction occurring there. Leather thongs for tying the pad to the saddle are attached near the front edge.

h. Accessories. Phillips pack saddlery includes several accessories as follows:

(1) *Blinds.* For use in blinding animals.

(2) *Spare hair bag.* An additional supply of curled hair for use in the saddle pads.

(3) *Tool roll.* A complete set of tools for adjusting the saddle pads (fig. 4).

(4) *Spare parts set.* A set of those parts most frequently needed for repairing the saddle (fig. 5).

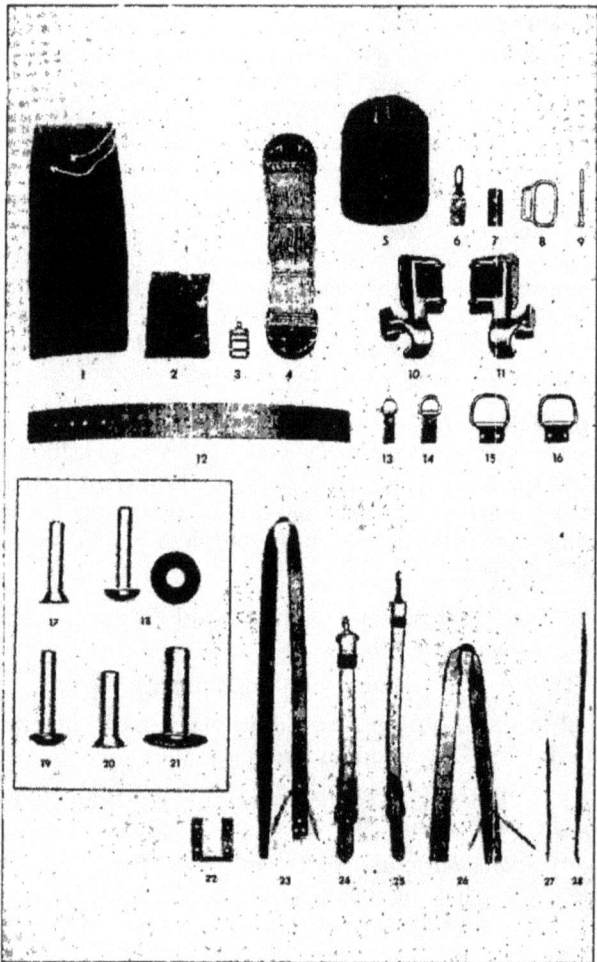

FIGURE 5. Spare parts, Phillips pack saddle, cargo type,
complete set.

Item No.	Name of part	Number in set
1.	Bag, spare parts, large	1
2.	Bag, spare parts, small	1
3.	Buckle, 1¾″, cinch, quick-release device	1
4.	Cinch, mohair, 20″	1
5.	Hair, spare (2-lb. bags)	2
6.	Hitchhook	1
7.	Hooks, lock-staple	2
8.	Loop, cinch, quick-release device	1
9.	Pins, bottom bar pocket, with spring	4
10.	Pocket, bottom bar, left	1
11.	Pocket, bottom bar, right	1
12.	Ribs, spring steel, frame	2
13.	Rings "D", .78″ x .65″, with clasp	2
14.	Rings "D", 1″ x .85″, with clasp	2
15.	Ring "D", 2″ x 1.5″, with clasp, left	1
16.	Ring "D", 2″ x 1.5″, with clasp, right	1
17.	Rivets, brass, ½″, countersunk head, No. 8	12
18.	Rivets, brass, ¾″, oval head, No. 8	24
19.	Rivets, steel, 3/16″ x 7/16″, button head	16
	Rivets, steel, 3/16″ x 11/16″, button head	8
20.	Rivets, steel, 3/16″ x 9/16″, countersunk head	3
	Rivets, steel, 3/16″ x 11/16″, countersunk head	3
21.	Rivets, steel, 1/4″ x 3/16″, wagon box head	3
22.	Staples, lock	2
23.	Straps, cinch	2
24.	Straps, holding, breeching	1
25.	Straps, hold-up, breeching	1
26.	Straps, quick-release	2
27.	Thongs, saddle, 1/4″ x 10″	12
28.	Thongs, saddle, 3/8″ x 24″	48

10. PHILLIPS CAVALRY PACK SADDLE. This saddle, with a few exceptions, fits the foregoing description of the cargo saddle. It has aluminum alloy arches and hanger bars and is equipped with a breast collar. The pads measure 22 by 19 inches. The saddle, being smaller than the cargo type, allows greater freedom of movement. Animals under full pack may work at increased gaits.

11. PHILLIPS CAVALRY PACK SADDLE, MODIFIED (CHINA SPECIAL). This saddle has the same outside dimensions as the cavalry pack saddle; however, its pads are much thicker. The arches and hanger bars are made of steel as in the cargo saddle. Equipped with a breast collar, it is suitable for carrying heavy, high-riding, howitzer loads used by pack artillery.

Section II. CARE AND PRESERVATION

12. GENERAL. Phillips pack saddlery should receive the same care and consideration as that given riding equipment. When in use all exposed parts and bearing surface should be cleaned daily, and the entire saddle completely dismantled and thoroughly cleaned once a week. *Pack saddles should not be stacked one on top of another, nor should men be allowed to sit upon them.*

a. Saddle pads. The curled hair and felt of the pads are impregnated with a moth-proof substance to protect them initially from moths. After the saddles have been in use and have absorbed sweat there is practically no danger from moths. This also applies to the woven pads. The felt facing (bearing surface) of the pads should be brushed clean and free of dirt and of salt from sweat. The grooming brush may be used for this purpose. Periodically, the pads should be detached from the frame for thorough cleaning of the leather backs. In arranging the equipment in camp, the lower edges of the pads should not be placed in contact with the ground. The footrests will prevent such contact provided reasonably firm ground is selected. In semipermanent bivouacs, or in wet weather, logs may be cut and the saddles placed thereon.

b. Saddle frames. No special precautions are required for the care of the saddle frames other than to keep all steel parts free of rust by painting. Painting also prevents reflection of light, an essential precaution to be taken in the combat zone.

c. Breeching. All leather should be kept clean and free of grit and dirt. The bearing surfaces of the breeching must be given careful attention. It is important that these parts be kept free of excess soap or oil. These substances collect dust and grit and thus cause the equipment to rub the animal.

d. Woven pads. In storage, woven pads should be protected from moths. Proper care should be taken to keep them smooth, soft, and free from grit at all times. When removed from the backs of pack animals, pads should be brushed to remove loose hair before it has had time to dry. The woven pads should receive a final brushing out several hours later to free them of salt from sweat. At this time, they should be worked and rolled between the hands to restore natural softness. Should the woven pads become excessively dirty, they may be washed in cool water with dissolved mild soap. They should not be wrung, but merely hung immediately to dry.

e. Care in shipping. If the pack saddles are to be shipped by boat, or over long distances, where they may be transferred from rail car to rail car, or truck to

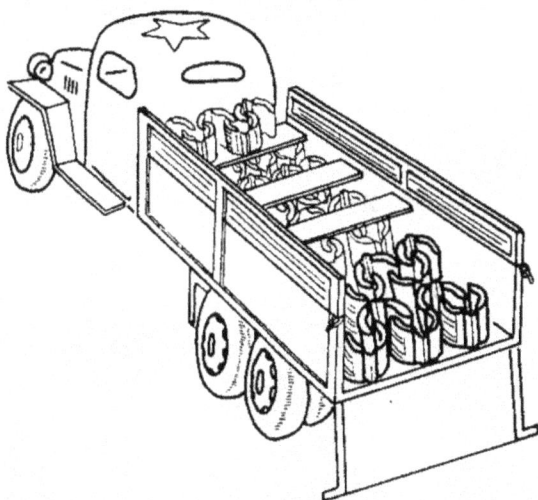

FIGURE 6. Pack saddles properly loaded in vehicle.

truck, they should be crated individually. However, when saddles are to be transported comparatively short distances by truck or rail they may be prepared as follows:

(1) Fold the woven pad once and lay it across the saddle. Place the breeching on top of the pad and secure both in place by fastening cinches tightly over the breeching. .

(2) Place the saddles on edge on the floor of the truck bed or railway car, dovetailing the rows so as to make the maximum use of floor space. Place boards on top of the first layer of saddles and add a second layer. If there is room, a third or fourth layer may be made in a similar manner (fig. 6).

Section III. FITTING AND ADJUSTING SADDLE

13. GENERAL.

a. Position of saddle. Too much stress cannot be laid on the proper positioning of the saddle and correct cinch adjustment. The forward edge of the saddle should be sufficiently in rear of the shoulder blades to allow the latter to operate freely. A clearance of 2 or 3 inches usually is sufficient. The saddle should not be canted to either side.

b. Adjustment of cinches.

(1) A pack saddle covers so great an area of moving surface that proper cinching is vital. Excessive binding of the front cinch may injure the back and sides, interfere with breathing, or cause cinch galls and swellings. The rear cinch must not be as tight as the front cinch, because the rear of the saddle covers the area of greatest motion, the flexible short-ribbed region, and the tender region over the kidneys. Since the hind legs are the propelling members, the hind-quarters move side-to-side and up-and-down. These movements must not be restricted by cinch pressure. There must be no interference with locomotion.

(2) The final adjustment of the cinch is made after the load is placed on the saddle. The exact amount of cinch pressure required can be determined only by experience. A safe rule to follow is to give the front cinch pressure sufficient only to hold the saddle in place. Usually one finger should pass easily between the front cinch and the animal's chest. The rear cinch should be cinched tight enough only to limit the rocking motions of the saddle and to help prevent the saddle from slipping forward. The rear cinch should never be so tight that the whole hand cannot be slipped under it. In testing cinch pressure, the finger or hand should be inserted from rear to front so that when it is withdrawn the hair will not be ruffled. Ruffled hair may cause galls.

(3) Excessive binding of the rear cinch will cause a pack animal to become exhausted more quickly.

c. Adjustment of breeching. With the saddle cinched in its proper position upon the animal, the breeching is adjusted as follows:

(1) Attach the hold-up straps to the top **D**-rings on the rear edge of the saddle pads; adjust them so that the croup piece will lay flat a few inches above the base of the tail. A general rule for clearance is one hand's breadth. Adjust the lead-up straps so that they hold the body piece inclined slightly downward to the rear. Attach the holding straps to the lower **D**-rings on the rear edges of the saddle pads. These straps should be adjusted to provide about 1 inch clearance between the body piece and the animal's buttocks. This completes the *approximate* adjustment of the breeching.

(2) The animal should be led at the walk to check the approximate adjustments. As one hind leg reaches its rearmost position in the stride, the body piece should bear flatly and snugly against it; at that moment, the holding strap on that side should be taut. Care should be exercised to insure against this adjustment being too tight, thus hindering the animal's natural gait; however,

FIGURE 7. Near packer attaching bits as off packer puts
woven pad in place.

if the holding strap is too loose, the breeching is use-
less.

(3) For traveling down long slopes, it may be necessary to tighten the holding straps two or three holes.
This will help materially in keeping the saddle off the
withers.

d. Adjustment of breast collar. The breast collar
as used on the cavalry pack saddle and the cavalry pack
saddle, modified, should be adjusted so that it is snug
when a front leg is fully extended, and bear against the
animal just above the point of the shoulder.

14. SADDLING. The method of saddling described
herein is applicable to the three previously described
types of pack saddles. An additional method of saddling is described in chapter 7, but normally it is used
only in herded pack trains.

a. The near packer snaps on the reins and bit, placing the bit in the animal's mouth, right to left, and

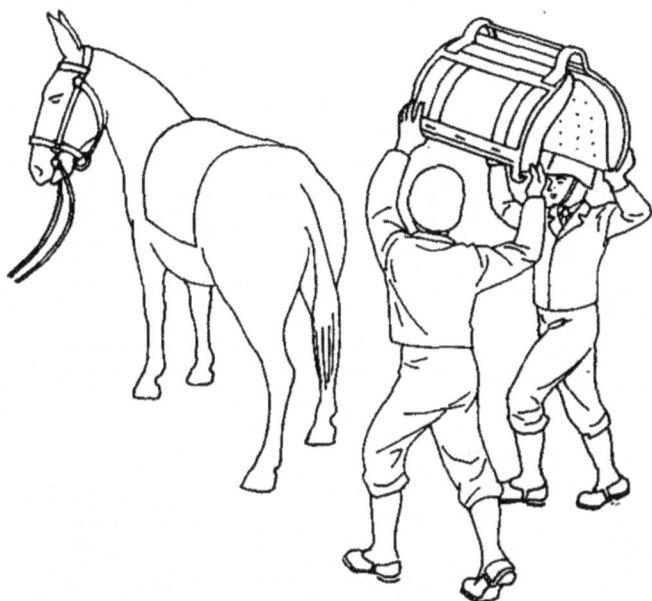

FIGURE 8. Packers lifting saddle prior to setting it on
animal's back.

loops the loose reins through the near cheek ring of the
halter bridle.

b. The off packer, after brushing off the animal's
back, sliding the woven pad, front to rear, places it in
position with its forward edge about a hand's breadth
in front of the rear edge of the shoulder blade (fig. 7).

c. Both packers grasp the pack saddle by the footrests, raise it high over the animal's croup, and lower it
gently into place (fig. 8). The saddle is placed so that
the front edge is 2 or 3 inches in rear of the shoulder
blade.

d. Using their right hands, both packers (near
packer to the rear, off packer to the front) grasp the
woven pad and the arch of the saddle with thumb and
forefinger. With their left hands, they grasp the left
footrests and raise the saddle slightly, simultaneously

pulling the woven pad upward to provide about 1 inch clearance between the woven pad and the animal's spine. This procedure is called "breaking the pad."

e. In lowering the saddle, packers exert outward pressure so as to properly seat it.

f. The near packer then secures the cinches in place while the off packer puts the breeching in place, lifts the animal's tail outside the breeching, and attaches the off holding strap.

g. The near packer attaches the near holding strap of the breeching.

h. Employing square bow knots, both packers tightly tie the thongs of the woven pad to the front D-rings of the saddle.

15. UNSADDLING. The method of unsaddling described herein is applicable to all three types of pack saddles. The additional method of unsaddling described in chapter 7, is used only in herded pack trains.

a. Packers untie the thongs of the woven pad. The cinches are unfastened and secured over the saddle. The holding straps are unsnapped and the breeching is inverted into place on top of the saddle. Grasping the saddle by the footrests, the packers lift the saddle upward and to the rear.

b. The near packer unsnaps the bits and reins and places them across the saddle.

c. The off packer removes the woven pad and places it on the saddle.

16. SADDLE PADS. Saddle pads are fitted and adjusted to *prevent* or *relieve* injuries. To prevent injuries, the saddle is adjusted initially to fit the animal, and thereafter as necessary to maintain the fit of the bearing surfaces of pads.

a. Making pad adjustments.

(1) Most adjustments of the saddle pads consist of either adding, removing, or rearranging the curled hair

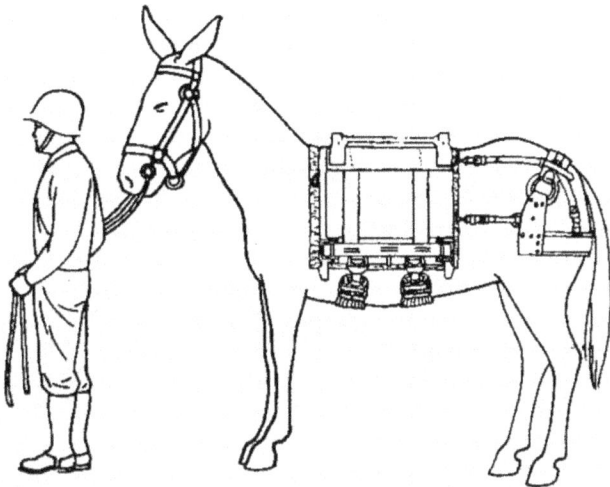

FIGURE 9. Phillips pack saddle, cargo, in proper position.

stuffing. Practically all adjustments are made through the hand-holes in the leather backs of the pads, thus conserving the original smooth bearing surfaces. For the work, pads are removed from the frame. All of the thongs are removed from the area to be adjusted.

(2) (a) The hair hook is used to remove hair from the pad. Hair always should be removed from the leather side of the pad so the bearing surface will retain its smooth contour.

(b) The stuffing rod is used to add hair to the pads. Prior to stuffing, hair should be well loosened, and only small amounts of it put in at a time. Hair is placed against the leather in the pad so as to avoid disturbing the established smooth contour of the bearing surface.

(c) The hammer is used to beat out the desired contour on the bearing surface of the pad. Care should be exercised to prevent striking the pad hard enough to break the woven felt surface. The pad should be struck only with the side of the hammer head, except in chambering a saddle, when pressure is applied by using the

35

ball peen to form the chamber. The hammer handle may be used to push thongs tightly into place as they are tied on the outside.

(d) The awl is used for replacing thongs in the pad.

(3) A chamber is a recess in the bearing surface of the pad intended to relieve pressure over an injury. When retying, thongs must be tied tightly so as to maintain the form of the chamber. It may be desirable to tie in an extra thong at the exact center of the chambered area if one is not already there. While the packmaster adjusts the size and shape of the chamber by pushing against the thongs with the hammer handle from the bearing side of the pad, an assistant ties the thongs on the back side. If no chamber is required, thongs, when being retied, should be tightened to the same tension as other thongs.

(4) In all pad adjustments, space must be allowed for the woven pad.

b. Initial fitting of the saddle. The new pack saddle will fit some animals sufficiently well that initial adjustments of the pads will not be necessary. In such cases the pads will shape to the animal's back after a small amount of use. However, the new pack saddle may fit many animals so poorly that pads should be adjusted so as to secure a more accurate individual fit before the saddle is used. The steps of fitting are as follows:

(1) The animals should be saddled without the woven pad and the fit observed from the front, sides, and rear. The front of the saddle should fit smoothly against the animal with no compression of the withers. The sides of the saddle should not be pushed outward excessively. The rear of the saddle should follow the body's natural curved lines without pinching. As viewed from the front and rear, the saddle pads should bear uniformly along the weight-bearing muscles of the back. Excessive bearing or compression at any one place is especially undesirable. The bottom bar should be horizontal or in-

clined slightly downward and forward; never downward toward the rear. Check the saddle for adjustment. If necessary, remove it, and make the adjustment.

(2) Powder the bearing surface of the animal's back with powdered chalk or flour. Place the saddle on the animal carefully and cinch lightly; no woven pad is used in this operation. Leave the saddle in place for a few minutes so as to insure complete transposition of the markings. Remove the saddle and examine the pads. The pads should have even powder marks over all the bearing surface. Where there is heavy powder marking, excessive hair must be removed. Where there is little or no powder marking, hair may have to be added.

(3) Saddle the animal with the woven pad in place, making sure the saddle is in the correct position. Cinch snugly and place an evenly-balanced full pay load on the saddle. Walk the animal until he shows signs of sweating and then remove the load and saddle. Remove the woven pad *very carefully* so as not to disturb sweat markings on the back. Check the back for pressure marks and spots showing little or no pressure. Excessive pressure is indicated by a definite area or areas on which the hair is considerably drier than on the surrounding areas, or which is free from any evidence of sweating. The hair over these areas will be either ruffled or compressed depending on the movement of the saddle. Wet surfaces on the back under the woven saddle pad indicates normal or little or no bearing from the saddle. Like the hair of the back, the woven pad will be driest over an area of excessive pressure. The character of sweating under the saddle is explained by the simple fact that pressure on the skin lessens the blood supply to the sweat glands and the excretion of sweat is diminished. If the pressure is very excessive, there will be no sweating. Locate the corresponding areas on the saddle and adjust them by removing or adding hair.

(4) Having been adjusted, the saddle now is ready for use on marches. It is essential that the packmaster ob-

serve new saddles very closely while they are being broken in, and that he promptly makes any adjustments which appear expedient.

c. Adjustments for maintaining the fit of the pads.

(1) Animals lose flesh when worked hard for long periods of time. As a result, contours of their backs change considerably. Saddle pads must be adjusted promptly to conform to these changes.

(2) As new saddle pads are broken in, the curled hair will pack down. It may be necessary, therefore, to add hair in order to maintain the proper contour of the pads.

(3) The packmaster must inspect his saddles and animals closely and promptly take all steps necessary to maintain the accurate fit of the saddle pads. In this manner, injuries will be minimized. Drivers and chiefs of sections also should observe closely and promptly report saddles which do not fit properly.

d. Adjustments to relieve injuries.

(1) The basic principles of relieving injuries due to the pack saddle or load are as follows:

(*a*) Find the cause of the injury.

(*b*) Remove the cause of the injury.

(*c*) Remove all pressure from the injury to allow it to heal.

(2) If saddlery is fitted properly, there is but small chance of its being the primary cause of injury to conditioned animals. Before the pads are adjusted, the cause of the injury is determined and a decision made as to whether or not it may be corrected by other means. Many injuries are relieved simply by correcting faulty positioning of the saddle, readjusting the breeching, changing cinch pressure, or balancing loads more carefully. Minor bruises ordinarily do not require removal of hair from the pads, nor is such corrective action necessary for a swelling that subsides, unless it is caused by a lump of hair in the pad. It often is possible to re-

lieve pressure by tightening a thong instead of removing hair.

(3) Basically, all pack saddle injuries may be relieved by reducing or removing pressure or friction from over the injured part; this is accomplished by chambering or making a recess in that part of the pad immediately over the injury. If an injury occurs where the saddle is too shallow to form a proper chamber, it may be necessary to build up the entire saddle in order to relieve the pressure at that point.

(4) When an injury occurs that requires chambering of the pad, the position of the chamber must be located accurately. This is best accomplished as follows:

(a) Sprinkle foot powder or flour over the injury.

(b) Place the saddle in its proper position without the woven pad.

(c) Cinch lightly.

(d) After sufficient time has elapsed to allow the powder to mark the pad, remove the saddle and detach the pad. The location of the injury will be indicated by powder mark on the pad.

(5) To chamber a saddle pad, loosen all thongs passing through the marked area so as to provide a chamber of at least 1 inch greater radius than the marked area. When loaded, the saddle will settle down considerably on the animal's back, making the center of the chamber, as marked by the powder, too low. Therefore, to compensate for this, the pad should be chambered about 1 inch higher than the center of the powder mark. Pull the leather slip out of the handhole nearest the marked area. Using the hair hook, remove hair as necessary, pulling small amounts at a time. Press in the chamber on the bearing surface or felt side and tie the thongs. Replace the pad in the frame and position the saddle on the animal to check the fitting. It may be necessary to use the powder marking system described in (4) above to insure accuracy of the check. The packmaster should inspect the injury frequently and carefully, making ad-

justments of the pad as they are needed, until the injury is completely healed.

(6) As soon as the injury is healed, the normal shape of the pad should be restored gradually.

(7) In restoring a pad's normal shape or in building up pads in the field, any suitable material such as sacking, grass, hay, or paper may be used if curled hair is not available. However, such expedients should be replaced as soon as possible.

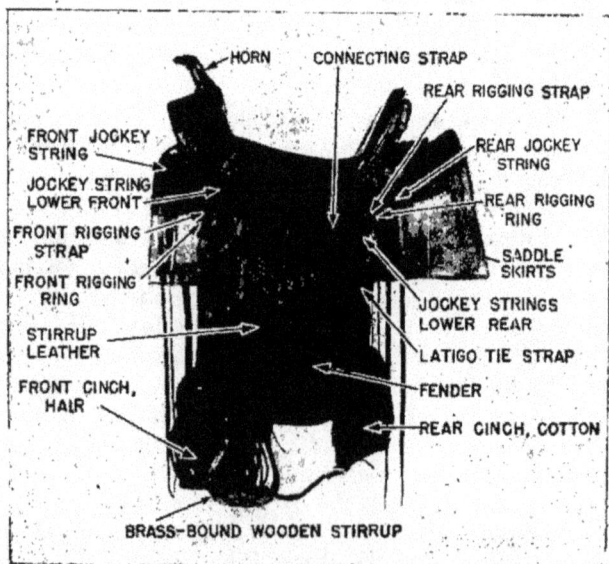

FIGURE 10. Packer's saddle, full-rigged.

CHAPTER 4
PACKER'S SADDLE, FULL-RIGGED

17. DESCRIPTION. The packer's saddle, full-rigged, is issued for riding purposes to personnel of pack units equipped with mules. It is a stock type saddle, full-rigged with two cinches (the front cinch of hair, the rear cinch of cotton). Cinches are connected by a strap and buckle to keep the rear cinch from working too far to the rear. The saddle's skirts are lined with sheepskin. Stirrup leathers have fenders attached, and are adjusted and secured with leather laces on the inner side. The stirrups are wooden and brassbound. The issue saddle blanket is used as a saddle pad.

18. POSITION OF SADDLE. The packer's s a d d l e should be placed on the animal's back so that the front ends of the saddle bars are approximately 2 or 3 inches in rear of the shoulder blade. If the saddle is allowed to ride too far to the front, cinch galls usually form rapidly under the front cinch.

19. ADJUSTMENT.

a. The stirrups of the packer's saddle should be so adjusted that when the rider stands in them, with the heels slightly down, there is about 1 inch clearance between his crotch and the seat of the saddle.

b. Cinches. See paragraph 13b.

c. Saddle blanket. The saddle blanket is folded as prescribed in paragraph 12, FM 25–5. It is placed on the animal's back with the front edge about two fingers' width in front of the rear of the shoulder blade.

20. SADDLING. Saddling is accomplished as prescribed in paragraph 15, FM 25–5, modified to include the cinching of the rear cinch.

CHAPTER 5
LASHED LOADS

Section I. PREPARATION OF CARGO FOR PACKING

21. GENERAL.

a. In pack units, supplies such as hay, grain, rations, and ammunition, may be wrapped in canvas cargo covers (mantas) and secured to the saddle with ropes. The shape, size, and weight of the cargo determine whether it will be packed on the saddle in one, two, or three bundles.

b. Cargo is wrapped for protection and support. Wrapping is dispensed with only when the cargo is of such nature as to not require protection or support. Loose, miscellaneous items of cargo may be packed in burlap sacks before being wrapped. For articles which have hard, smooth surfaces, wrapping greatly aids in keeping the ropes of hitches in place on the load.

c. The equipment used in preparing cargo for packing is the 6 by 6-foot canvas cargo cover (manta) and the lair rope. The lair rope is 30 feet long and 3/8 inch thick. It has an eye spliced in one end. When used for slinging loads, it is known as a sling rope.

22. WRAPPING CARGO.

a. To wrap an item, the cargo cover (manta) is spread evenly on the ground, lair rope under one corner, and the cargo placed diagonally across the center. The packer standing at the corner of the cargo cover (opposite the lair rope) (fig. 11), picks up the corner nearest to him, draws it snugly over the cargo, tucking the end under the cargo if necessary, places one or both knees on top, reaches over, grasps the opposite corner, folds it

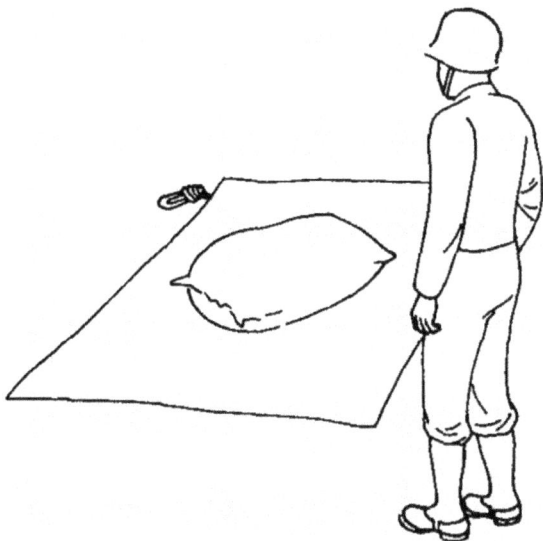

FIGURE 11. Cargo on cargo cover preparatory to wrapping.

FIGURE 12. Placing doubled portion of cargo cover in
position.

in, and brings the doubled portion over the top of the
bundle so that the edge lays just past the center (fig.
12). He places his left knee on top of the bundle to hold
the cargo cover in position. The cargo cover then' is

FIGURE 13. Tightening lengthwise loop.

crimped in at the right end and the flap brought over
the bundle. The packer next places his right knee on
the flap to hold it in position, and the operation is repeated for the left end, the latter being folded under
so that the folded edge will come to position near the
middle and on top. Placing his knee upon the flap to
hold it in position, he takes the lair rope, forms a loop
by passing the end through the eye, places it lengthwise
around the middle of the bundle, and draws it taut so
that the eye comes near the top of the left end (fig. 13).
He then takes one half hitch around the bundle at the
end nearest the eye of the rope, another in the middle

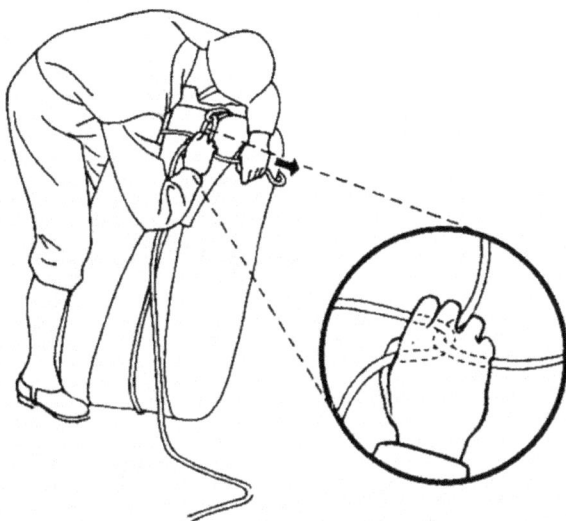

FIGURE 14. Adjusting first half hitch.

of the bundle, and a third one around the other end. Finally he wraps the remainder of the rope around the bundle lengthwise and ties it on top, using a *sliding clove hitch* (figs. 14, 15, 16, 17, 18).

b. Bed rolls, cord wood, coils of lash and sling ropes, and similar loads do not need the protection of the cargo cover. Two packers, working together, may lair such loads as follows:

(1) A lair rope is spread on the ground so that its center portion forms a "Z" (fig. 19).

(2) The cargo is placed upon the "Z."

(3) Each packer passes his end of the rope over the bundle and through the bight (loop) on the opposite side, and takes up the slack by pulling the rope toward him. The ends are then passed lengthwise around the bundle, and after the slack has been taken out of the rope, each is secured on top with two half hitches (fig. 20).

FIGURE 15. Tightening first half hitch.

c. Extra Phillips pack saddles should be prepared as pack loads as follows: the saddle to be transported is disassembled; saddle pads, woven pad, and breeching must be wrapped as two bundles. The frame is placed on top of the saddle on which the load is to be packed. The two bundles then are lashed on top of the frame as side loads. By placing the frames one on top of the other, two saddles can be carried.

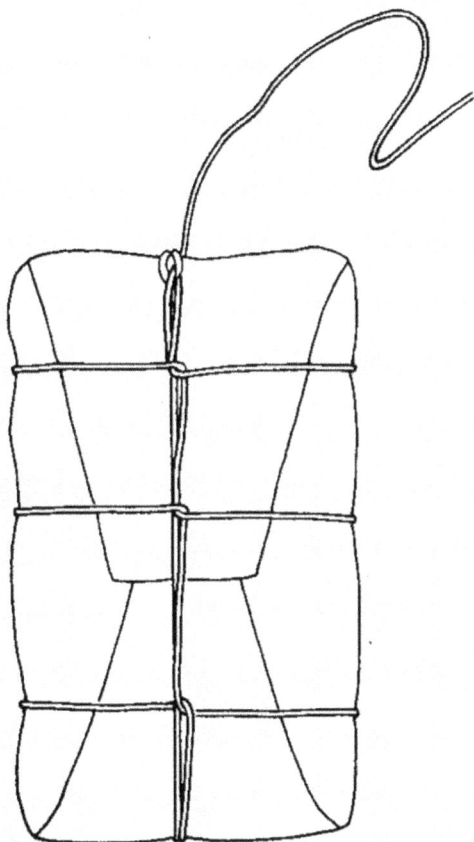

FIGURE 16. Half hitches in position.

FIGURE 17. Tying off lair rope with sliding clove hitch.

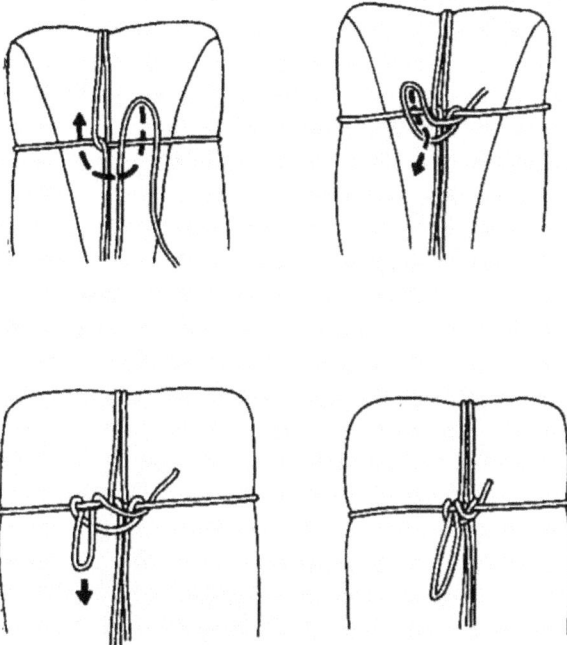

FIGURE 18. The sliding clove hitch.

FIGURE 19. "Z" rope formation for lairing cargo.

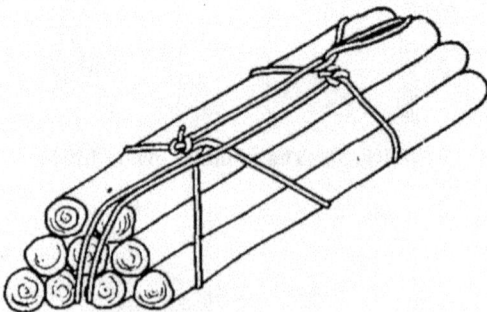

FIGURE 20. Laired cargo.

Section II. SLINGS

23. GENERAL.

a. If the total load is composed of two side loads, or two side loads and one top load, it is necessary to sling these loads before lashing in order to hold the loads in place while the hitch is being formed. Wrapped side loads are placed on the saddle with the folds of the cargo cover to the outside and with eye splice of the lair rope always to the packer's left. This method insures the maximum protection against damage from rain, brush, sand, and snags. The loads carried in all slings should be settled in place as follows:

(1) Both packers lift upward and outward on their loads until the loads rest flat against the saddle.

(2) Each packer then grasps the top of his side load and assists in settling the load by pulling down on it. This method applies only to evenly balanced side loads.

(3) When one load is heavier than the other, the heavier must be slung higher on the saddle to obtain the required transverse balance.

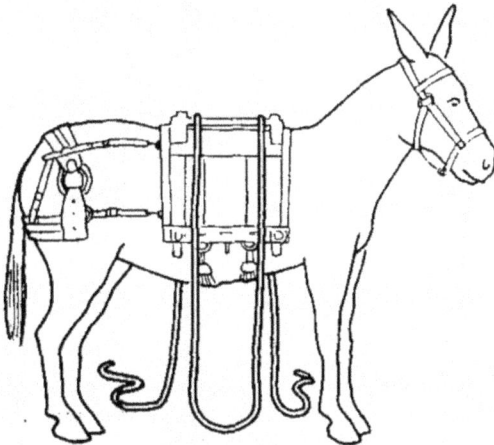

FIGURE 21. Adjustment of sling rope for cross-tie sling.

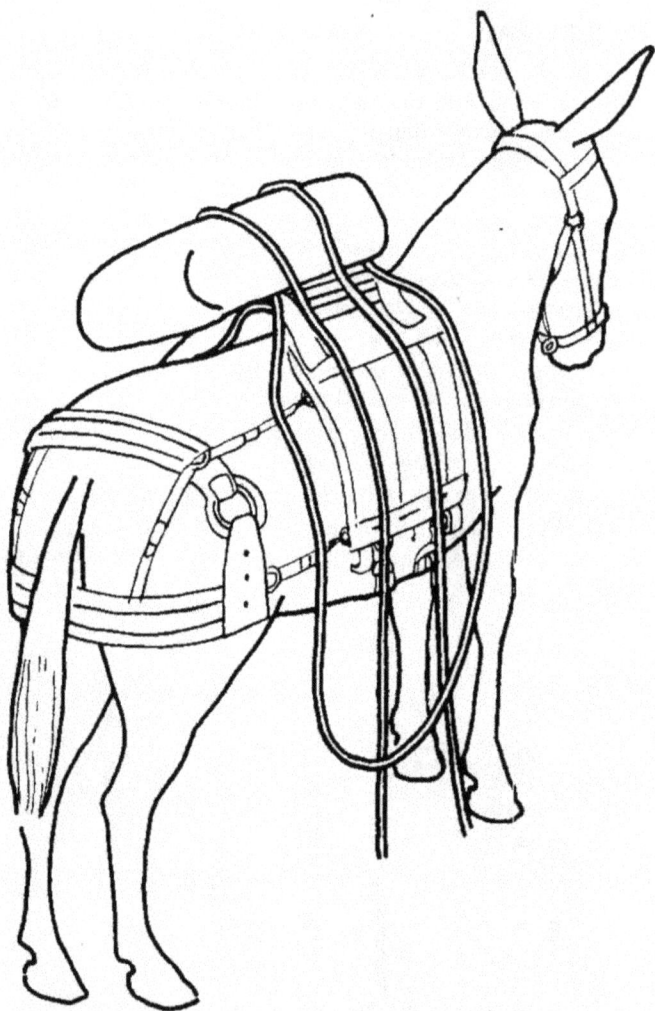

FIGURE 22. Ends of sling rope thrown over near side load
to off side of saddle.

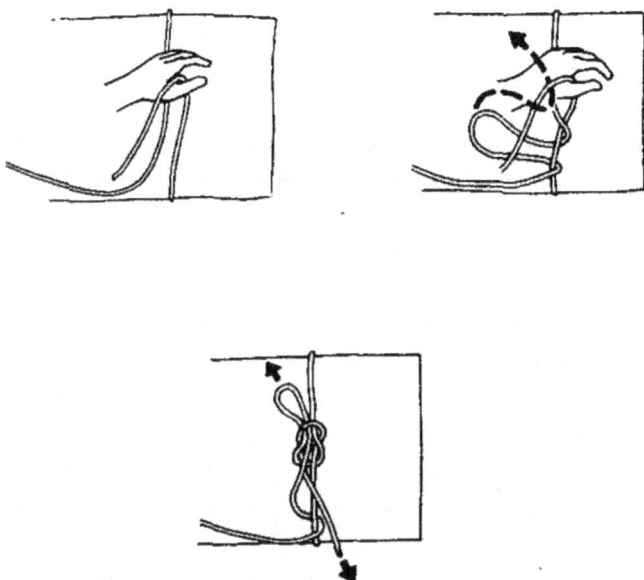

FIGURE 23. Square bowknot.

(4) Loads should be positioned so as to have a slight excess of weight forward.

b. There are many types of slings which may be used; however, those listed herein are considered most practicable and adaptable. These slings are: the cross-tie sling, the double-tie sling, and the single-tie sling.

24. CROSS-TIE SLING.

a. The off packer gets the sling rope (lair rope) and, holding the center, throws both ends of it over the saddle, forming a loop on his side. He adjusts the ropes so that they lay across the saddle approximately 1 foot apart, and the loop on his side just touches the ground (fig. 21).

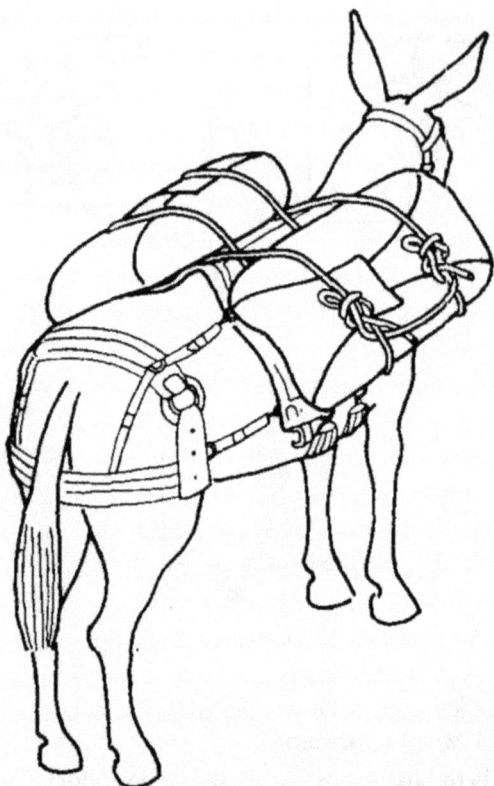

FIGURE 24. Cross-tie sling complete.

b. The near packer gets his load and places it high on the saddle, centered, and in a horizontal position. He then throws the ends back over his load to the off side.

c. The off packer spreads the end ropes 1 foot apart and lays the front and rear ropes of the loop off the saddle (fig. 22). He then puts his load on the saddle, end ropes underneath.

d. Supporting the load with his left forearm, he takes the front rope of the loop and flips it over the

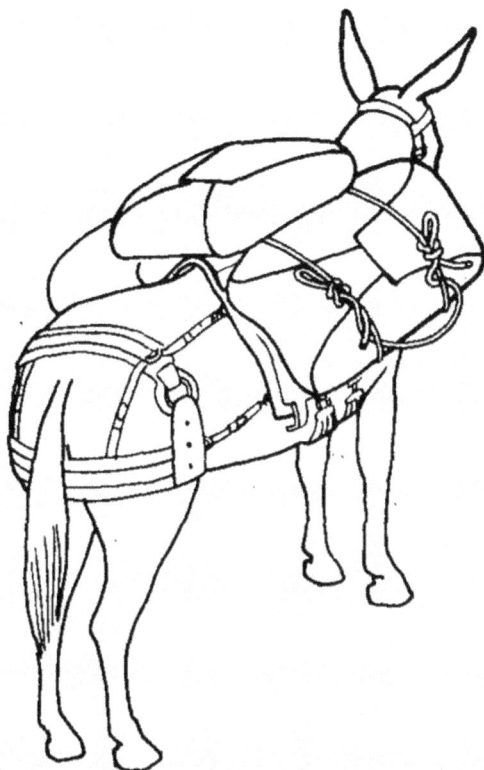

FIGURE 25. Cross-tie sling with third load in place.

load. Making sure that both ropes are taut, he ties the front end rope to the front rope of the loop with a square bowknot (fig. 23). He then moves his left forearm to the rear of the load and ties the rear end rope to the rear rope of the loop in the same manner as he did the front (fig. 24).

e. All excess ropes are tucked away as the near packer goes after the lash rope.

f. If the triple load is to be used, the third bundle is placed on top of the two side loads already slung (fig. 25).

25. CROSS-TIE SLING (ALTERNATE METHOD OF TYING).

a. The near side packer gets the sling rope, uncoils it, doubles it at its center, and passes the loop thus formed across the saddle.

b. The near packer adjusts the running (free) ends so that they are approximately 1 foot from the ground, and passes any excess rope to the off side. Both packers spread the ropes so they lay about 1 foot apart across the saddle.

c. The off packer passes the looped end of the rope across between the ropes to his partner who adjusts it so that it also hangs approximately 1 foot from the ground.

d. With the two loops now on the off side, the outside ropes are laid just off the saddle both front and rear, and the middle two ropes are spaced approximately 1 foot apart (fig. 26).

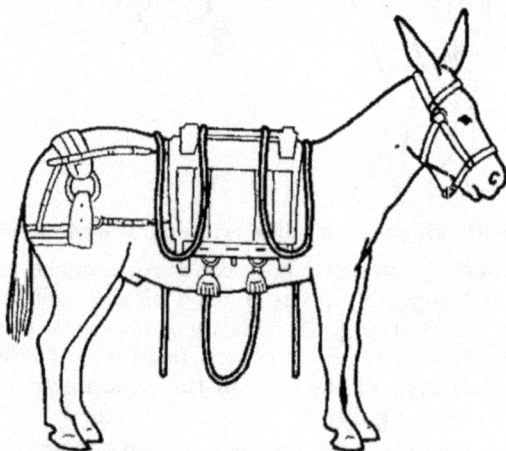

FIGURE 26. Adjustment of sling rope for cross-tie sling (alternate method).

e. The off packer now places his load high on the saddle, centering it in a horizontal position. Assisted by the near packer, he then lifts the outside ropes of the sling over the front and rear ends of his side load.

f. The near packer calls "Loop over," and tosses the loop of the sling over the off side load (fig. 27). He then raises and positions his side load.

FIGURE 27. Position of loop over the off side load (alternate method, cross-tie sling).

g. The off packer, now calls "Loop over," and tosses the loop back over the near side load.

h. The near packer, supporting the load with his left forearm, removes all slack and secures the rear running rope to the rear rope of the loop with a square bowknot. He then moves to the front of the load and repeats the operation on the two front ropes, tucking away all loose ends on top of the load. The sling now is complete.

i. While the near side packer is tucking away loose ends, the off packer checks the load for balance, and prepares the lash rope for the hitch to be used.

26. DOUBLE-TIE SLING. To form the double-tie sling, two packers work together as follows:

a. The off packer doubles a sling rope near its center and passes the loop thus formed over the saddle to the near side, letting the ends hang down on his side so that they just touch the ground. He then spreads the ropes so they are about 1 foot apart (fig. 28).

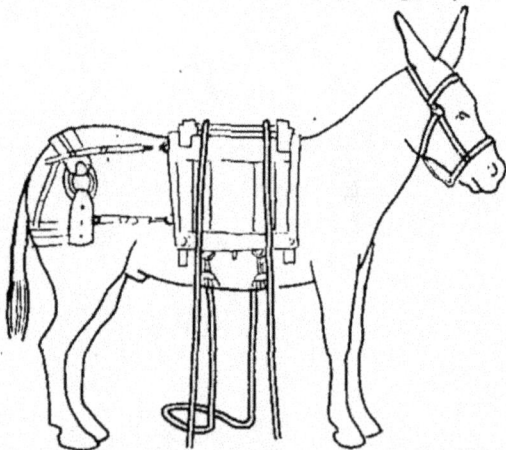

FIGURE 28. Adjustment of sling rope for double-tie sling.

b. The near packer then places his load well up on the saddle, resting it on the sling rope with a flat side in contact with the saddle, and holds it in position with his left forearm. The off packer then picks up his load, places it on the saddle so that it overlaps the near side load by about 4 inches, and holds it in position with his left forearm. The near packer then passes the loop of the rope over the top of the load. Upon receiving it,

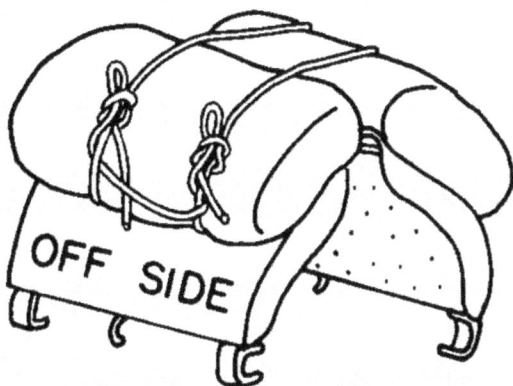

FIGURE 29. Double-tie sling complete.

the off packer takes out the slack and ties (in turn) a square bowknot in both the front and rear ropes (fig. 29).

   c. The load then is settled in place. While the off side packer secures the loose ends of the sling, the near side packer prepares the lash rope for the hitch to be used.

   d. If a triple load is to be used, the third load is placed on top of the two loads already slung.

27. SINGLE-TIE SLING.

   a. The off side packer gets the sling rope, uncoils it, retains both ends, and throws the remainder over to the near side of the saddle. He adjusts the rear rope so that it just touches the ground.

   b. The near side packer adjusts the front rope of the sling so that the loop on his side is about 6 inches off the ground. For small loads, loops may be somewhat smaller (fig. 30).

   c. The ropes of the sling are separated so that they lay about 1 foot apart.

   d. Both packers now place their loads high on the saddle, centering them in a horizontal position.

59

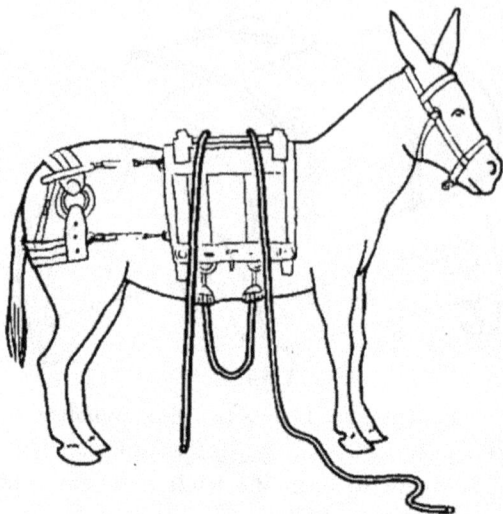

FIGURE 30. Adjustment of sling rope for single-tie sling.

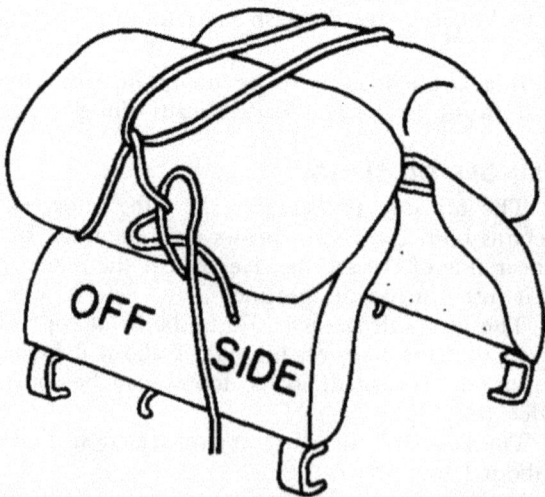

FIGURE 31. Single-tie sling complete except for square
bowknot.

e. The near packer calls "Loop over" and passes the loop to the off packer.

f. The off packer passes the rear rope of the sling through the loop, pulls it snug, and ties it firmly to the front rope of the sling with a square bowknot (fig. 31).

g. Loose ends are tucked away, and the load is checked for balance and settled in place.

Section III. RESTRAINT OF PACK ANIMALS WHILE PACKING

28. GENERAL. When properly trained, most pack animals will stand quietly while being packed. However, some are so restless that to pack them efficiently presents a real problem. There are several methods for keeping restless animals quiet while they are being packed. Three such methods are as follows:

a. By tying the animals close (not over 6 inches) to a picket line, post, or other object on which they cannot hurt themselves.

b. By requiring a third packer to stand to the front of the animal and distract him by scratching his head, rattling the bits, or patting his nose. *Actions must be gentle.* In no event should the animal be punished by a twitch or by being eared-down.

c. By using the blind. This method is a last resort. The packer, standing on the near side of the animal, grasps the blind at the button end, and passes it fully over the animal's neck, allowing the half of the blind with thong to hang down to the off side. He then brings the crown of the blind well to the front and, with his right hand, passes it, right to left, over the animal's ears, being careful not to startle the animal. The blind then is secured by the thong provided for that purpose. To remove the blind, the foregoing operation is reversed. Headshy mules occasionally are found to resist application of the blind. In this case the blind may be put in place by sliding it over the nose of the animal, on up over his eyes, and securing it in place with the sliding knot under the jaws. The thong then is passed over the animal's neck and secured to the button on the near side. *Under no circumstances should an animal be moved, no matter how slightly, while blinded.*

Section IV. HITCHES

29. GENERAL. A hitch is a formation of lash rope used to hold cargo securely on the pack saddle. The standard lash rope is ½-inch in diameter, 60 feet long, and has an eye spliced in one end. There are many different types of hitches which may be used. Most are based on the same principles. No one hitch, however, is suitable for all loads. Therefore, it is desirable to train personnel of pack units in the use of several hitches to enable them to tie securely any load which they may be called upon to handle.

30. FORMATION OF HITCHES.

a. Rules governing the formation of all hitches are as follows:

(1) Form the hitch rapidly. Three minutes is sufficient time for qualified packers to complete any hitch.

(2) Check to make sure the load is balanced *before* and *after* tightening the hitch.

(3) Keep ropes away from the animal's feet and legs.

(4) Work quietly to avoid confusion.

(5) Make all parts of the hitch as tight as possible.

(6) Make certain that no ropes are fouled on the saddle or breeching.

b. The hitches used in pack units and considered best for the specific loads indicated are as follows:

(1) *Squaw hitch.* Single loads.

(2) *Phillips cargo hitch.* Double box loads.

(3) *Single diamond hitch.* Double loads of normal size and shape.

(4) *Double hitch.* Odd-shaped double loads.

(5) *Double diamond hitch*. Triple loads.

(6) *Basket hitch*. Odd-shaped loads. Double loads may be packed by one packer. Effects a low center of gravity.

c. Other satisfactory hitches are as follows:

(1) *Sweeten diamond hitch*. Loads of normal size and shape. Also triple loads.

(2) *Nagle hitch*. Two side loads or a single load.

31. TIGHTENING A LASH ROPE.

a. Almost all standing and quarter ropes of the hitches are tightened by the packer to remove slack. This sometimes is known as "hitting a rope." This should always be done with the greatest amount of

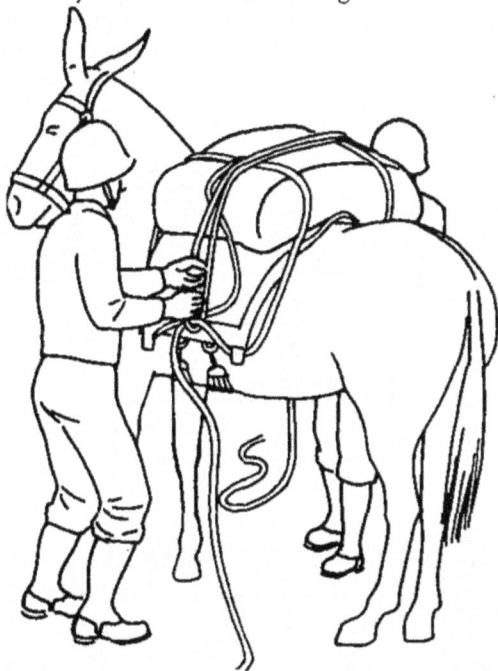

FIGURE 32. The near packer preparing to tighten ("hit") the standing rope.

strength to assure that the rope is as tight as possible. The standing rope is that part of the lash rope which binds the center or main part of the load to the saddle. The quarter ropes are those parts of the lash rope which bind the ends of the loads to the footrests.

b. The procedure for taking slack out of a lash rope on a load is as follows: the packer who is to tighten a standing rope or quarter rope on his side of the hitch grasps the rope, with both hands, near the footrest or hitch hook, places his feet together, directly below his hands (fig. 32), and then calls "Take slack." His partner grasps, with his right hand, the running end of the rope to be tightened, and at the same time places his left hand on the load to steady the saddle; his feet are slightly separated, giving him a balanced position (fig. 33). He now calls "Go." The packer tightening the rope gives two quick strong pulls outward and upward, us-

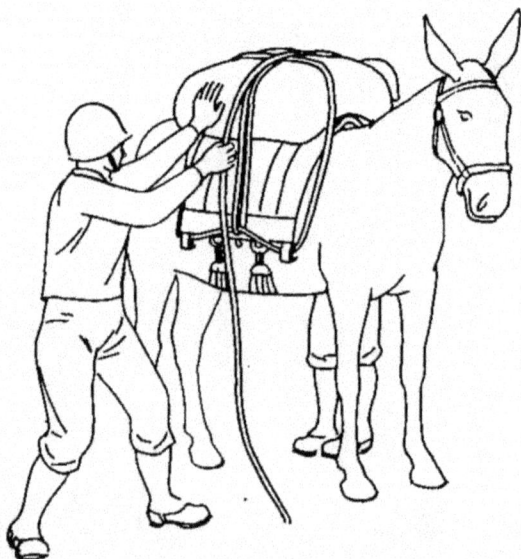

FIGURE 33. The off packer preparing to take slack, (squaw hitch).

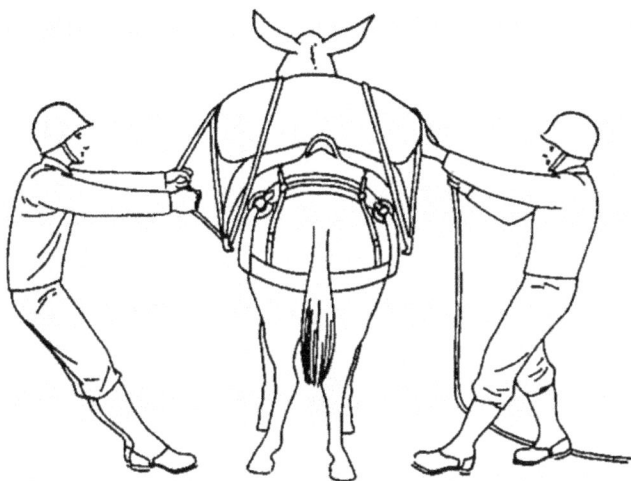

FIGURE 34. The off packer taking slack as near packer "hits," (squaw hitch).

ing the entire weight of his body (fig. 34). Simultaneously, the other packer, taking the slack from the running end of the rope, pulls with his right hand and pushes with his left, being careful not to lose any of the tension developed by the two strong pulls.

32. SQUAW HITCH. The squaw hitch is used when a single load is to be lashed to the saddle. In forming the hitch, two packers perform identical operations simultaneously.

a. Packers place the load on the saddle, center it, and balance it. On soft loads, each packer grasps the end of the load on his side and both pull down simultaneously to settle the load in place. While the near packer steadies the load, the off packer checks the load and obtains the lash rope. Retaining the center of the rope, he throws the remainder out to the rear, freeing it of kinks and knots. Each packer now grasps the rope near the center and lays it across the load, allowing the running (free) ends of the rope to drop down opposite

65

the hitch hooks and the remainder to lay to his right
(fig. 35).

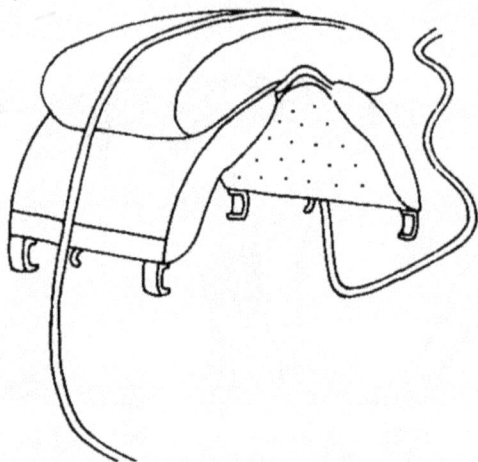

FIGURE 35. The first step in formation of squaw hitch.

b. Each packer, with his left hand (palm down),
grasps his running rope about 7 feet from the load and
places his left hand on the standing rope where it passes
over the end of the load. Passing the right hand to the
left of the running rope which is at his right, he grasps
the standing rope at a point just below the load (fig.
36). Being careful to keep the running rope about opposite the hitch hook, he spreads his hands, forcing the
half hitch thus formed toward the top of the load (fig.
37). Keeping the half hitch in place with his left hand,
he guides the rope in his right hand (taking up slack
as he goes), down over the right side of the load, through
the footrests, and back over the left side of the load to
the half hitch (fig. 38). With the right hand, he grasps
the running rope below the half hitch and pulls out
slack. The near packer now calls "Ready," and the off
packer, when ready, calls "Down," and both (with both
hands) pull down simultaneously on their running
ropes, taking slack, and seating the load (fig. 39). The

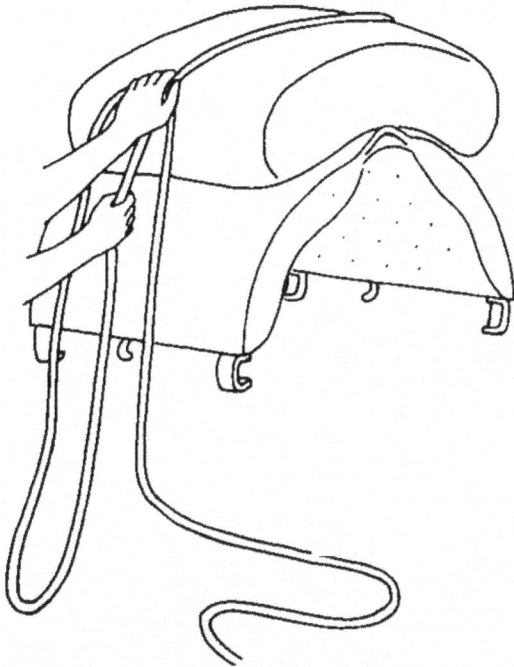

FIGURE 36. Forming loop with left hand and grasping standing rope with right hand (squaw hitch).

packers grasp the running rope just below the saddle and pull hard outward and upward, removing some slack from the standing rope between the two foot-rests. Holding the running rope taut by binding it against the standing rope with the left hand, he picks up the remainder of the rope with his right hand and passes it across the load to the opposite packer, keeping it to the right. Each packer takes the rope just passed to him, takes up excess slack, and makes a temporary bight under the standing rope, right to left. (For some loads or when ropes are wet, it may be desirable to make this temporary bight from *left* to *right* in order to avoid binding when slack is taken from the standing ropes.) The near packer moves far enough to the rear to check

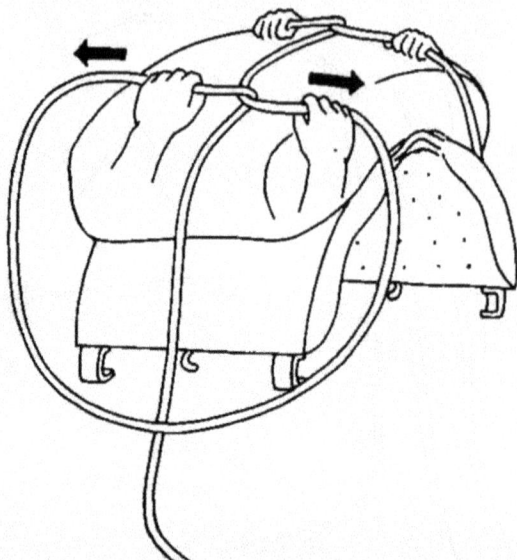

FIGURE 37. Spreading hands and forcing half hitch to top of load (squaw hitch).

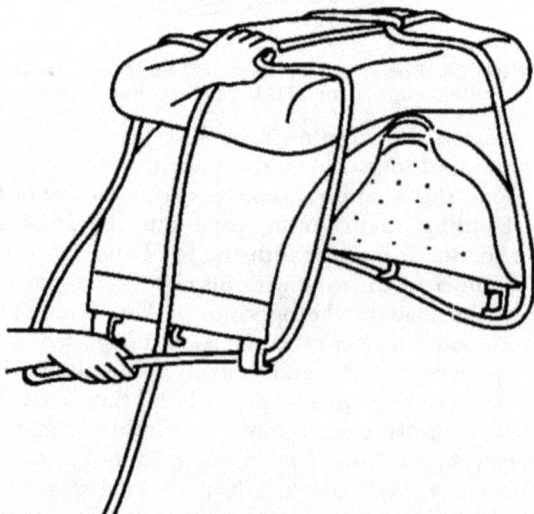

FIGURE 38. The half hitches of squaw hitch in place.

FIGURE 39. Removing slack from quarter ropes and seating load (squaw hitch).

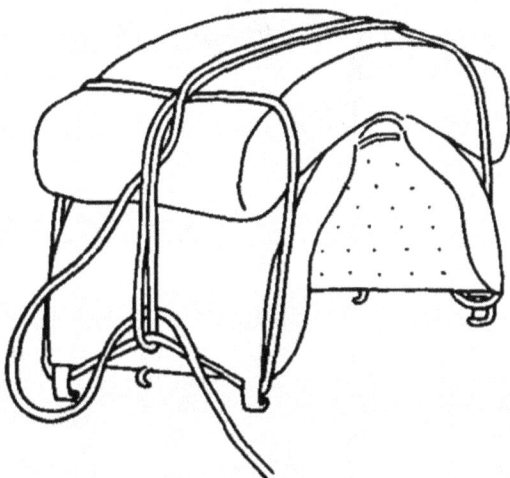

FIGURE 40. The permanent bight in place (squaw hitch).

the balance of the load. If the load is not balanced, the packer at the higher side tightens ("hits") his standing rope first. When both standing ropes have been tightened, a permanent bight is put in on both sides from right to left (fig. 40).

c. The hitch now is ready to be hobbled on both sides. A hobble is a formation of the running end of a lash rope used to tighten further the quarter ropes after the hitch has been formed. Each packer, after bighting the running rope from right to left, forms a loop in this rope and passes it under the left quarter rope from inside to outside. The part of this loop which extends past the left quarter rope should be as small as possible

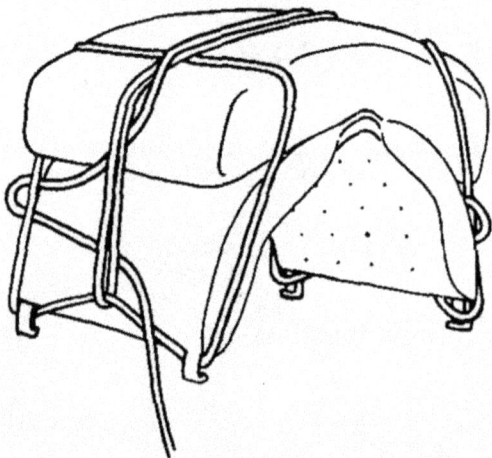

FIGURE 41. Beginning of formation of hobble (squaw hitch).

(fig. 41). Holding the loop with his left hand, each packer, with his right hand, forms a second loop in the remainder of the running rope. He passes this loop over and back under the right quarter rope and then through the loop held in the left hand (fig. 42). Grasping the second loop with both hands, he pulls hard either to the front (near side) or rear (off side). Keep-

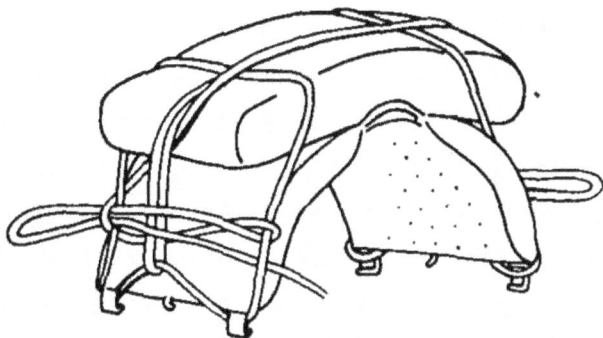

FIGURE 42. Formation of hobble (squaw hitch).

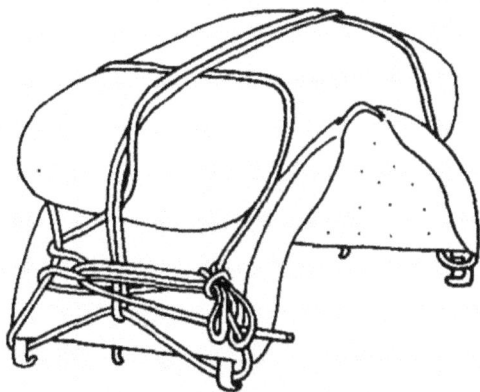

FIGURE 43. Squaw hitch complete.

ing all slack, he then steps either to the rear (near side) or front (off side), places his inside foot against the footrest, and pulls out all remaining slack. The hobble is tied on the right quarter rope with a sliding clove hitch, or two half hitches, and all loose ends are tucked away (fig. 43).

d. For extremely short or round loads, a modification of the squaw hitch may be made to keep the quarter ropes on the load. This is accomplished by forming a

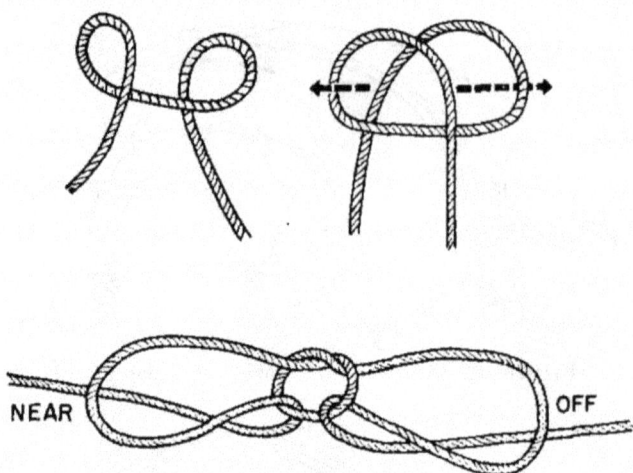

FIGURE 44. Formation of butterfly knot.

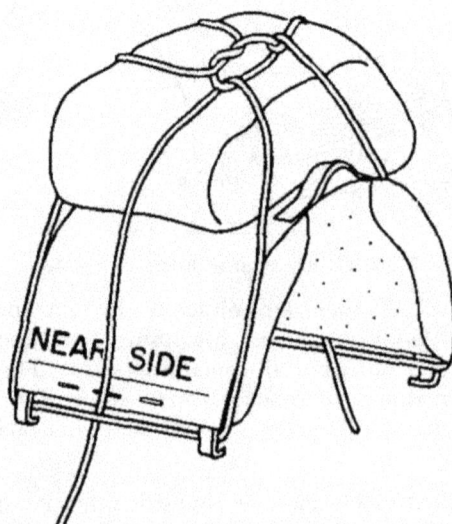

FIGURE 45. Formation of the quarter ropes of squaw hitch
using butterfly knot.

butterfly knot (fig. 44) at the center of the lash rope, placing the knot at the top center of the load, and forming the quarter ropes by enlarging the two loops of the butterfly knot. Running ends, centered at the hitch hook, hang below the loop formed (fig. 45). The hitch then is secured as described in b and c above.

33. SINGLE DIAMOND HITCH. The single diamond hitch is formed and tightened by two packers working together as follows:

a. If the cross-tie sling described in paragraph 24 is used, the off packer finishes tying the sling while the near packer gets the lash rope and, from the rear, checks the balance of the load. Adjustments are made as necessary.

b. The near packer, facing the saddle on his side, keeps the eye of the lash rope in his right hand and drops the remainder of the rope to the ground on his left. With his right hand, he then hooks the eye on the hitch hook and, sliding his right hand along the rope, forms a small loop which he passes over the cargo to the off packer, keeping the running rope to his left with his left hand (fig. 46).

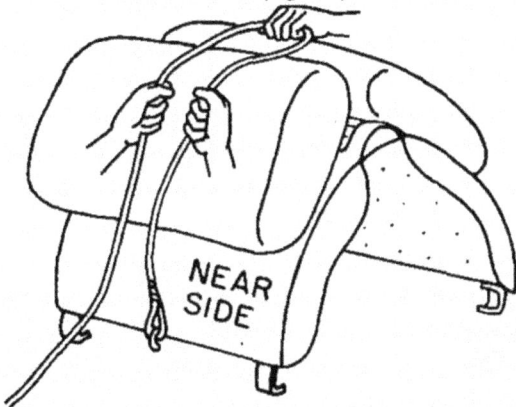

FIGURE 46. Handing small loop over cargo to off packer (single diamond hitch).

c. The off packer grasps the loop and, pulling any needed slack from the front or running rope, hooks the standing rope under the hitch hook, rear to front. The near packer pulls the running rope snug and bights it under the standing rope, rear to front, leaving enough loop hanging from this bight to pass under the front footrest and to form his rear quarter rope (fig. 47).

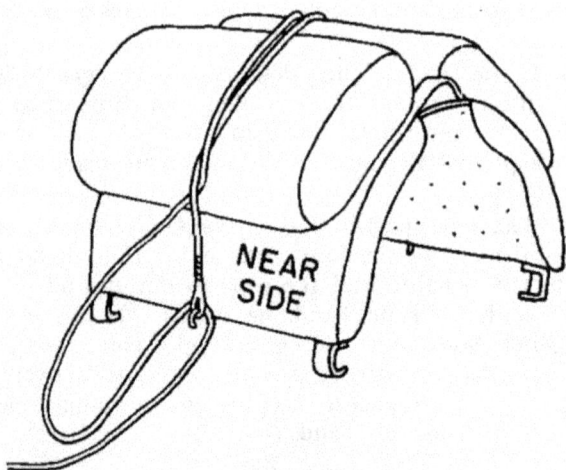

FIGURE 47. Bight in place with hanging loop (single diamond hitch).

d. The near packer, keeping the running rope to his left, lays a second loop over the load. The off packer grasps this loop, pulls it down between the standing ropes, and spreads it front and rear, thus forming the off quarter ropes. He adjusts the quarter ropes so they cross under the standing ropes between the loads. He takes any needed slack from the front quarter or running rope (figs. 48 and 49).

e. The near packer takes out his bight, and then takes slack, while the off packer tightens the front standing rope. Keeping the front standing rope tight, the near packer passes it under the rear standing rope,

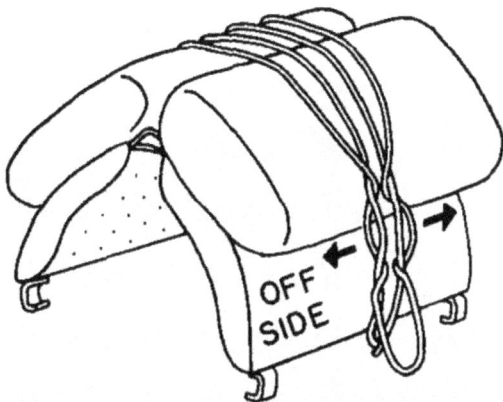

FIGURE 48. Second loop placed between standing ropes
(single diamond hitch).

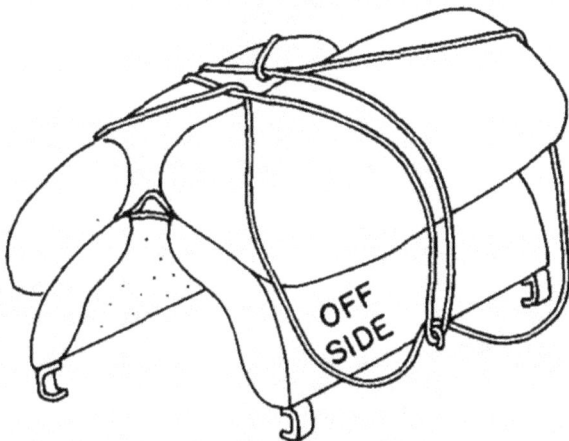

FIGURE 49. Second loop spread to form quarter ropes
(single diamond hitch).

rear to front, binding it between the load and the standing rope. He then passes it under the footrests, front
to rear, and finally over the top rear corner (near side)
of the load. He presses it tightly against the load while
the off packer takes up the slack in his rear quarter
rope.

FIGURE 50. Off packer taking slack while near packer pulls
down on load (single diamond hitch).

f. The near packer holds down the rear of his load
with both hands as the off packer further tightens the
ropes with a strong pull to the rear (fig. 50). While the
off packer takes slack, the near packer tightens his rear
quarter rope.

g. The off packer passes the running rope down over
the rear corner of his load, under the footrests, rear to
front, and then up over the front corner of the off
load, taking out all slack as he goes. He presses the
running rope tightly against his load until the near
packer has taken up the slack in his front quarter rope.

h. With both hands, the off packer holds down the
front of his load while the near packer tightens the rope
with a strong pull to the front. The off packer tightens
his front quarter rope while the near packer takes slack.

i. The near packer passes the running rope under the front footrest, front to rear, and then forms the near side hobble as follows:

(1) Having passed the front quarter rope under the front footrest, he holds it with his left hand.

(2) With his right hand (back outward), he loops the lash rope around the rear quarter rope, outside to inside, the loop coming out under the lash rope and leaving the end of the lash rope hanging down outside the footrest.

(3) He next grasps the loop with his left hand and moves it forward outside the standing ropes, then around the front quarter rope from the inside (fig. 51A).

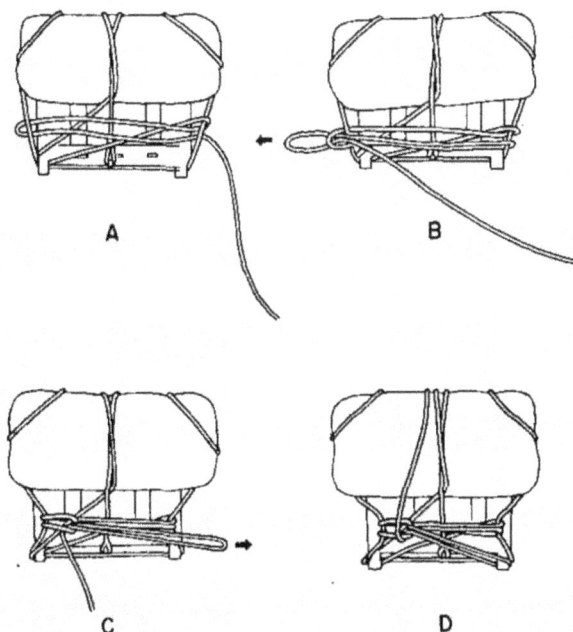

FIGURE 51. Tying the near side hobble (single diamond hitch).

(4) He forms second loop in the end of the lash rope hanging outside the rear footrest and passes it through the loop in his left hand. He removes slack by pulling the second loop to the front (fig. 51B) and, stepping to the rear, pulls out all remaining slack (fig. 51B and C). (If necessary, he places his inside foot against the footrest to obtain additional leverage.)

(5) He then hooks the loop under the rear footrest and passes the running end of the rope across the load to the off side (fig. 51). He tightens ("hits") this rope before the off packer forms his hobble.

j. The off side hobble is formed as in the squaw hitch (figs. 41, 42, 43).

34. DOUBLE HITCH. The double hitch is formed and tightened by two packers working together as follows:

a. If the method described in paragraph 24 is used, the off packer finishes tying the sling while the near packer gets the lash rope and, from the rear, checks the balance of the load.

b. When necessary adjustments in balance have been made, the near packer, facing the saddle on his side, keeps the eye of the lash rope in his right hand and drops the remainder of the loosened rope on the ground to his left.

c. The near packer, with his right hand, hooks the eye on the hitch hook and, sliding his hand along the rope, forms a small loop which he passes over the load to the off packer, keeping the running rope to his left with his left hand.

d. The off packer grasps the loop and, pulling any needed slack from the front or running rope, hooks the standing rope under the hitch hook, rear to front. The near packer pulls the running rope snug and bights it under the standing rope, rear to front, leaving sufficient loop hanging from this bight to pass under the front footrest and to form his rear quarter rope.

e. The near packer lays a second loop over the load, keeping the running rope to his left. The off packer grasps this loop and pulls it, front to rear, under the standing ropes between the loads. Taking his slack from the front rope, the off packer enlarges the loop, front and rear, to form the quarter ropes (figs. 52 and 53).

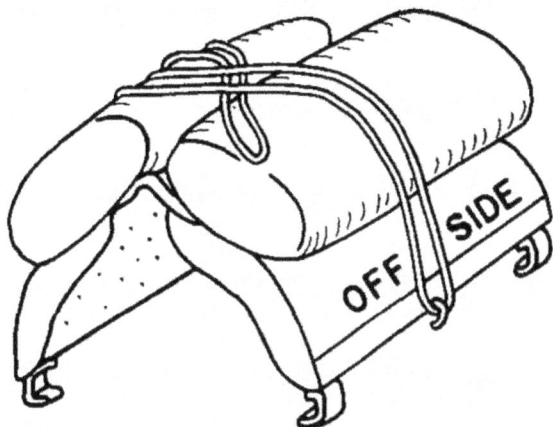

FIGURE 52. Second loop pulled under both standing ropes (double hitch).

f. The hitch now is tightened and hobbled as in the single diamond hitch.

35. PHILLIPS CARGO HITCH.

a. If the method described in paragraph 24 is used, the off packer finishes tying the sling while the near packer gets the lash rope and, from the rear, checks the balance of the load.

b. When necessary adjustments in balance have been made, the near packer, facing the saddle on his side, keeps the eye of the lash rope in his right hand and drops the remainder of the loosened rope to the ground on his left.

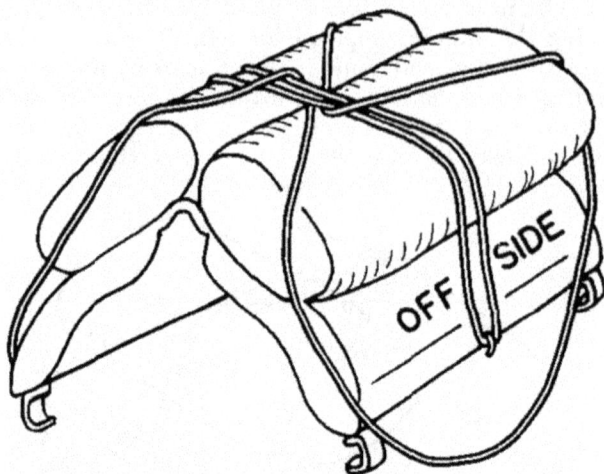

FIGURE 53. Second loop adjusted to form quarter ropes (double hitch).

c. The near packer, with his right hand, hooks the eye of the lash rope on the hitch hook and, sliding his right hand along the rope, forms a small loop which he passes over the load to the off packer, keeping the running rope to his left with his left hand.

d. The off packer grasps the loop end, pulling any needed slack from the front or running rope, hooks the standing rope under the hitch hook, rear to front. The near packer pulls the running rope snug and bights it under the standing rope, rear to front, leaving enough loop hanging from this bight to pass under the front footrest and to form his rear quarter rope.

e. The near packer lays a second loop over the load, keeping the running rope to his left. At this time, the near packer takes out his bight and takes slack while the off packer tightens ("hits") the front standing rope. Keeping the front standing rope tight, the near packer passes it under the rear standing rope, rear to front, binding it between the load and the standing rope, passes it under the footrests, front to rear, and up over the top rear corner of the near load.

f. The off packer next takes the rear rope of the second loop and, standing by the off front footrest, takes slack while the near packer tightens ("hits") his rear quarter rope. The off packer then guides it over the top front corner of the off load, through the footrests, front to rear, and up over the top rear corner of the load, retaining all slack as he goes (fig. 54).

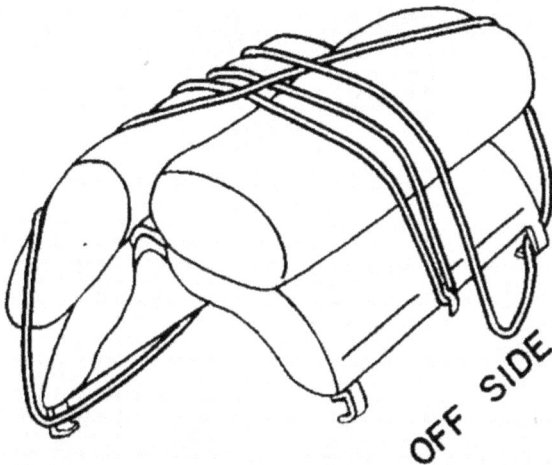

FIGURE 54. Formation of front quarter ropes (Phillips cargo hitch).

g. The near packer takes slack on the lash rope while the off packer tightens ("hits") his rear quarter rope. The near packer then guides it down over the top front corner of the near load and under the front footrest, front to rear (fig. 55). The hitch then is hobbled as in the single diamond.

36. DOUBLE DIAMOND HITCH. The double diamond hitch, used only for triple loads, is formed and tightened as follows:

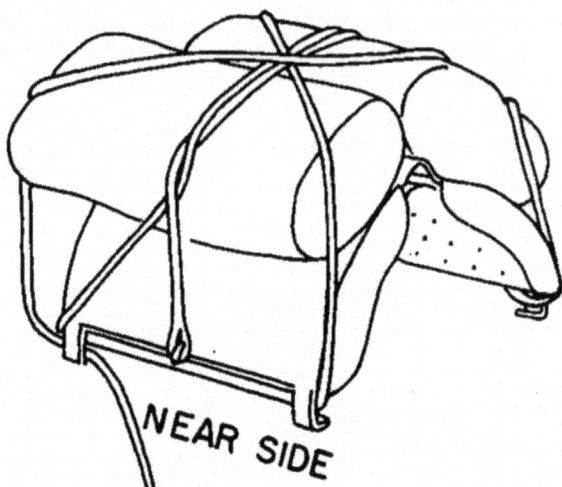

FIGURE 55. Phillips cargo hitch complete except for hob-
bling.

a. If the method described in paragraph 24 is used,
the off packer finishes tying the sling while the near
packer gets the lash rope and, from the rear, checks the
balance of the load.

b. When necessary adjustments in balance have been
made, the near packer, facing the saddle on his side,
keeps the eye of the lash rope in his right hand and drops
the remainder of the loosened rope to the ground on
his left. With his right hand, he hooks the eye on the
hitch hook, and sliding his right hand along the rope,
forms a small loop which he passes over the load to the
off packer, keeping the running rope to his left with
his left hand (fig. 56).

c. The off packer grasps the loop and, pulling any
needed slack from the front or running rope, hooks
the standing rope under the hitch hook, rear to front.
The near packer pulls the running rope snug and bights
it under the standing rope, right to left. He leaves sufficient loop hanging from this bight to pass under his
front footrest and to form his rear quarter rope.

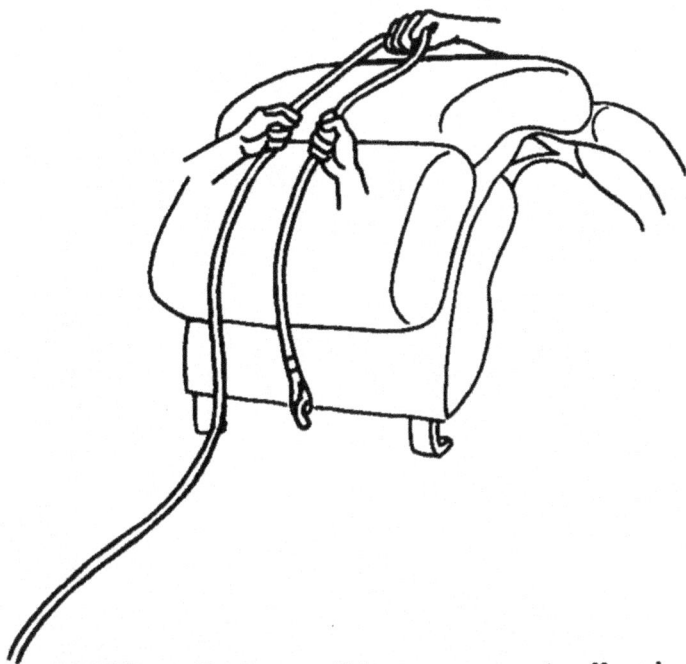

FIGURE 56. Handing small loop over cargo to off packer
(double diamond hitch).

d. The near packer lays a second loop over the load,
keeping the running rope to his left. The off packer
grasps this loop and enlarges it so that it hangs approximately 2 feet below the bottom bar.

e. Both near and off packers grasp the ropes of the
second loop between the side and top loads (one side of
the loop in each hand) and pass them down between
and under the standing ropes, inside to outside. The
near packer says "Ready," and the off packer, when
ready, says "Go" (fig. 57).

f. At the word "Go," both packers, with both hands,
simultaneously pull sharply upward and somewhat outward, thus pulling from under the standing ropes that
portion of the second loop that goes over the top load

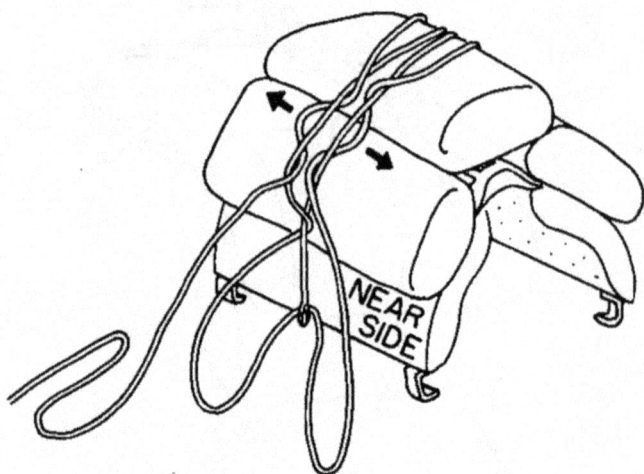

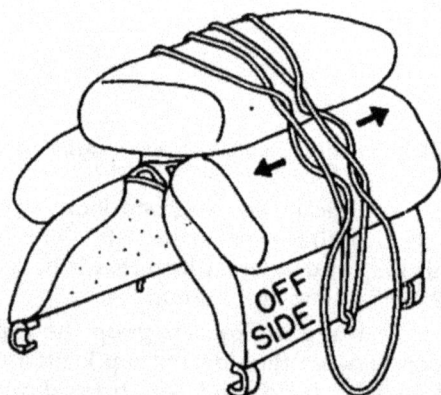

FIGURE 57. Passing ropes down between and under the standing ropes, inside to outside (double diamond hitch).

(fig. 58); thus the end ropes for the top load are formed. The ropes are then released.

g. Both packers place their front and rear quarter ropes to the front and rear of their respective side loads (fig. 59).

FIGURE 58. Pulling second loop out to form end ropes for top load (double diamond hitch).

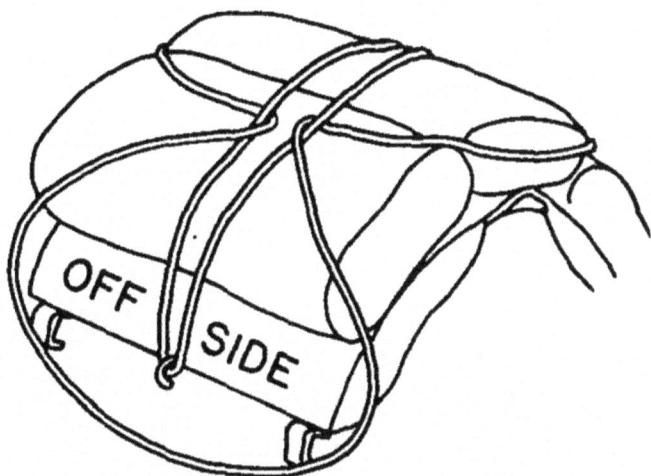

FIGURE 59. End ropes in place (double diamond hitch).

h. The near packer takes out his bight and takes slack while the off packer tightens ("hits") the front standing rope. Keeping the front standing rope taut, the near packer passes it under the rear standing rope, right to left, binding it between the load and the standing rope. He then passes it under both footrests, front to rear, and up over the top rear corner of the near load. Holding the rope in place on the load with his left hand, he grasps with his right hand the running rope that lays between the loads and pulls out all slack. With both hands, he grasps this rope up close to the standing rope, places his inside foot against the rear footrest, and pulls hard to the rear, removing all slack (fig. 60). The near packer now passes the running rope

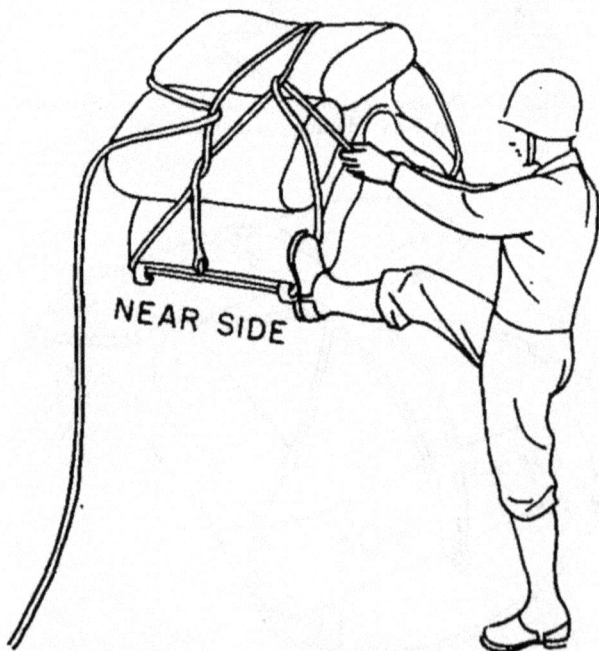

NEAR SIDE

FIGURE 60. Near packer grasping running rope with both hands to remove all slack (double diamond hitch).

around the rear end of the top load. At the same time, the off packer removes all slack by pulling on his rear quarter rope.

i. The off packer next grasps the rear quarter rope with both hands and, placing his inside foot against the rear footrest, pulls out all remaining slack. He then passes the running rope over the top rear corner of the off load, under the footrests, rear to front, and up over the top front corner of the off load. At the same time, the near packer grasps the running rope, which hangs down on the front corner of the near load, and pulls out excess slack.

j. With both hands, the off packer now grasps the rope that lays between the loads and, placing his inside foot against the front footrest, pulls hard to the front, removing all slack (fig. 61). He guides the running rope around the front end of the top load while the near packer pulls out the remaining slack.

k. With both hands, the near packer grasps the running rope up close to the standing rope, places his

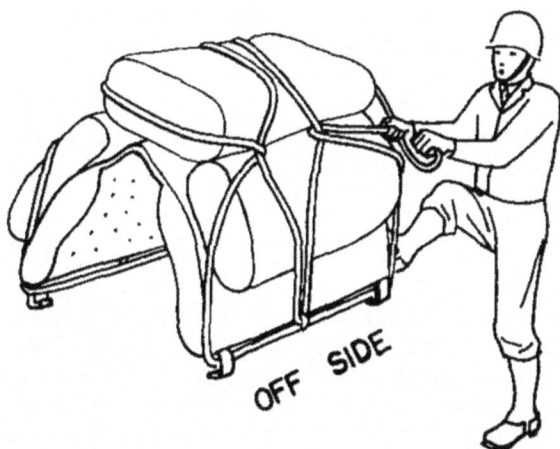

FIGURE 61. Off packer grasping rope laying between the loads and removing all slack (double diamond hitch).

inside foot against the front footrest, and pulls out all slack. He now guides the running rope over the top front corner of the near load and under the front footrest. Holding the rope with his left hand, he forms the near side hobble as in the single diamond hitch (fig. 51). He then passes the running rope across the load where it is hobbled on the off side as in the squaw hitch (figs. 41, 42, 43). The hitch now is complete. (If there is insufficient lash rope to form the off side hobble, it is made by using the halter shank.)

37. BASKET HITCH. The basket hitch, used for double loads, is formed and tightened by two packers working together as follows:

a. To form this hitch, it is necessary to secure the lash rope to the saddle before starting to pack because a sling rope is not used. It is accomplished in the following manner:

(1) The near packer passes approximately 2 feet of the center of the lash rope through the front arch of the saddle, rear to front, passes the ends through the loop formed, and pulls the loop tight. He separates the two running ends of the rope, one to the near and one to the off side.

(2) Each packer passes his end of the rope through the rear arch, rear to front, over the hanger bar, and under the loop thus formed.

(3) The running rope is carried toward the front of the saddle, sufficient slack being pulled to cause the bottom of the loop to hang even with the bottom bar (fig. 62).

b. Each side load generally is placed high upon the saddle in a vertical position and centered. The loop formed by the lash rope between the arches now is placed around the load and tightened, front to rear. All possible slack is taken out by pulling the running end of the rope, which at this time still is laying to the front of the saddle (fig. 63). The running rope then is

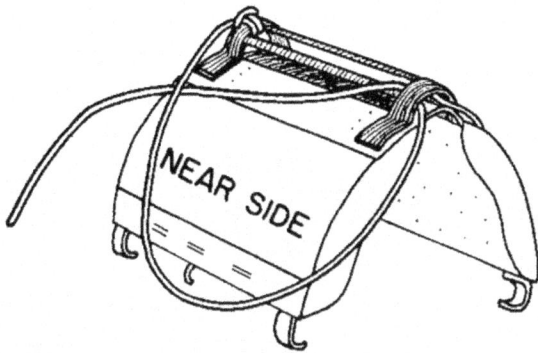

FIGURE 62. The initial formation of lash rope for basket hitch.

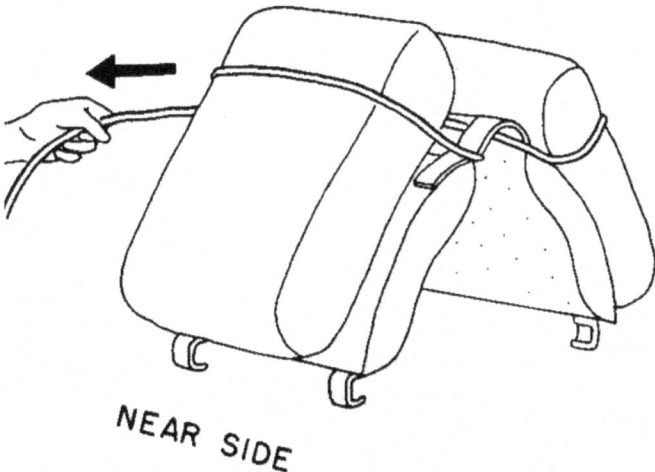

FIGURE 63. Taking out slack by pulling on running rope (basket hitch).

brought down and out under the center of the lower end of the load.

c. The running rope is doubled, running end to the left, and the loop passed at the center of the load under and over the standing rope. Sufficient slack then is taken in the loop to form quarter ropes which will pass

under the footrests. The running end remains centered
and beneath the loop formed (fig. 64).

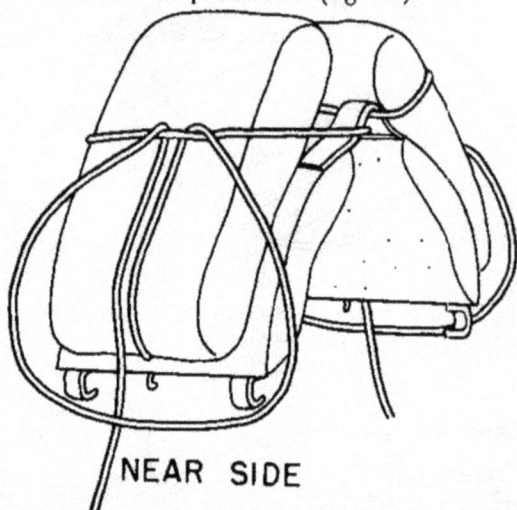

NEAR SIDE

**FIGURE 64. Formation of quarter ropes and running rope
(basket hitch).**

d. In forming the quarter ropes, the rope to the right
is pulled taut and passed under the right footrest, under
the left footrest, and all the slack then is removed by
pulling down hard on the running rope (fig. 65). The
running end hangs loosely behind the standing rope
between rear and front footrests. Further tension is
obtained by grasping this running end just below the
standing rope and giving a sharp upward pull on the
running end. Maintaining this tension, the running
ropes are passed across the center of the load to the opposite packer as in the squaw hitch.

e. The hitch generally is hobbled on both sides. The
running end of the rope received from the opposite
side is passed under the hitch hook, right to left, and
formed into the standard hobble as used on the squaw
hitch (figs. 41, 42, 43), making sure that the hobble
is kept below the bottom edges of the loads (fig. 66).

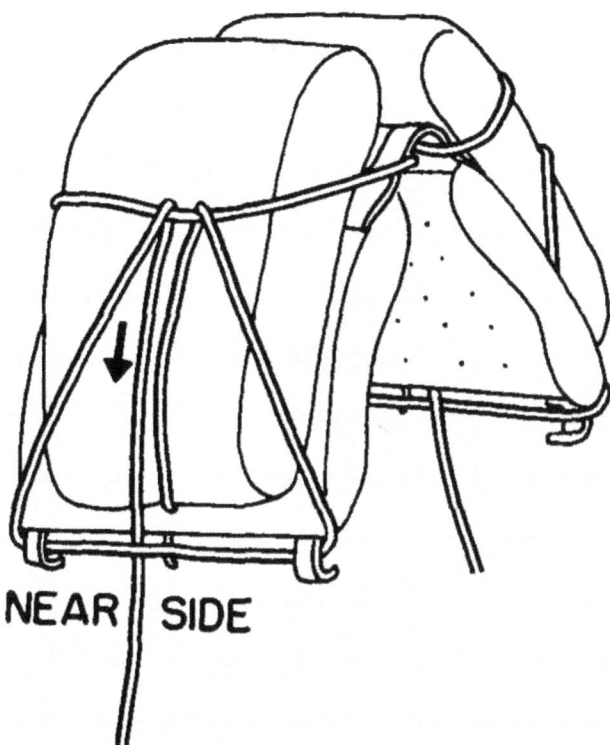

NEAR SIDE

FIGURE 65. Pulling on running rope to take up slack
(basket hitch).

f. For large or long loads, when insufficient rope
remains to form the hobble, the hitch is secured as follows:

(1) Form bowline knot in the end of the off running
rope and pass it across the load.

(2) Pass the running rope (near side) through the
loop of the bowline knot, pull down strongly, and secure the rope (fig. 67).

38. ADJUSTABLE LASH ROPE. This lash, designed
for use with the Sweeten diamond hitch or as a load
cinch, consists of two 4-inch rings, to each of which are

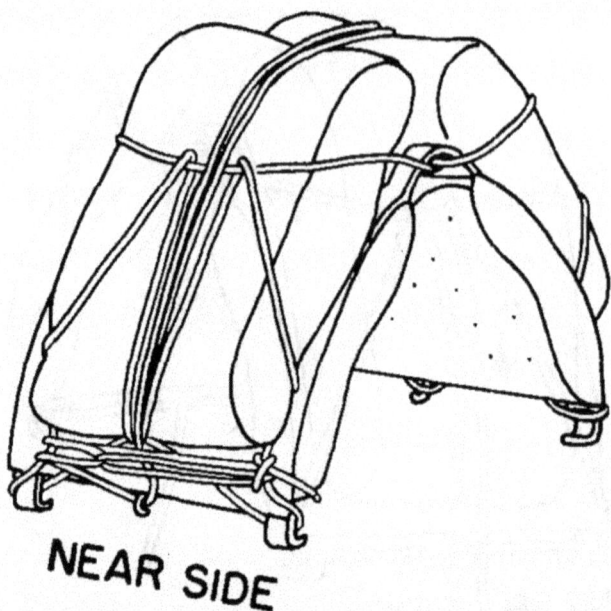

NEAR SIDE

FIGURE 66. Complete formation of basket hitch.

spliced 25 feet of ½-inch manila rope. The rings are connected by 9 feet of ½-inch rope, one end of which is spliced to one ring, the other end being connected to the same or opposite ring by a metal snap.

a. For ordinary loads, the rope with the snap is passed through the opposite ring and snapped to the first ring, thus providing a load cinch approximately 4½ feet long.

b. For very small loads, the rope with the snap is passed through both rings and snapped to the opposite ring, thus providing a load cinch approximately 3 feet long (fig. 68).

39. SWEETEN DIAMOND HITCH. The Sweeten diamond hitch is of special value to pack trains. It requires a special lash rope (par. 38), but the latter is useful when

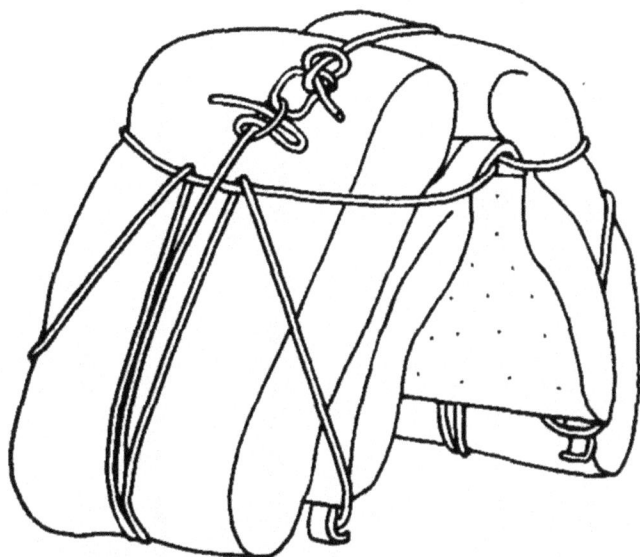

NEAR SIDE

FIGURE 67. The basket hitch tied off without a hobble.

a load cinch is required. Basically, this hitch is similar to the single diamond hitch and therefore is used for the same purposes. The hitch is formed and tightened by two packers working together, performing similar operations simultaneously, each on his own side, as follows:

a. If the sling described in paragraph 24 is used, the near packer gets the lash rope while the off packer finishes tying the sling. While in the rear, the near packer checks the balance of the load. Working together, the near and off packers place the load cinch part of the lash rope over the center of the load so that the 4-inch rings are equally distant from the top center of the load. If the size of the load and the type of pack saddle used will permit, the load cinch part should be

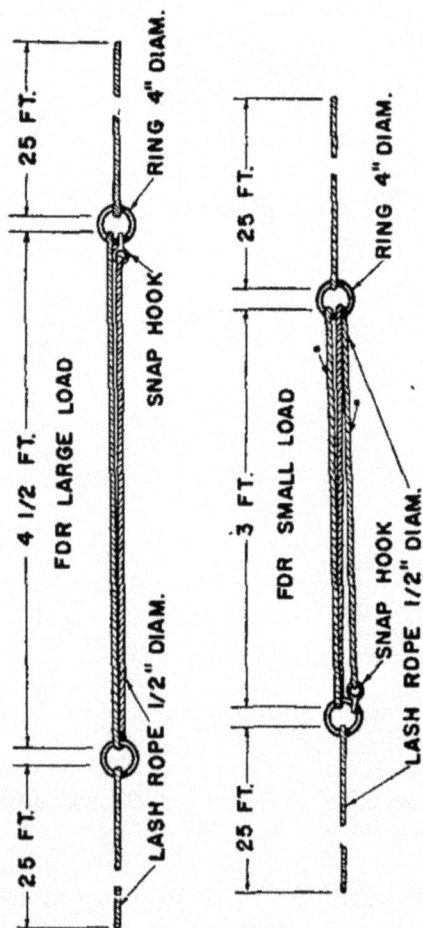

25 FT.

RING 4" DIAM.

25 FT.

SNAP HOOK

FOR LARGE LOAD

4 1/2 FT.

LASH ROPE 1/2" DIAM.

25 FT.

RING 4" DIAM.

25 FT.

SNAP HOOK

FOR SMALL LOAD

3 FT.

LASH ROPE 1/2" DIAM.

25 FT.

NOTE— *IN FORMING DIAMOND, LASH ROPE TO BE THROWN
AROUND OUTSIDE ROPES ONLY.

ADJUSTABLE CENTER ROPE IS 9 FT. LONG

FIGURE 68. Adjustable lash rope.

adjusted so that the rings are just below the side loads
(fig. 69).

b. Each packer, holding the ring on his side in position with his left hand, grasps his running rope in his
right hand and passes it, right to left, under both footrests. Still holding the ring with his left hand, he passes

94

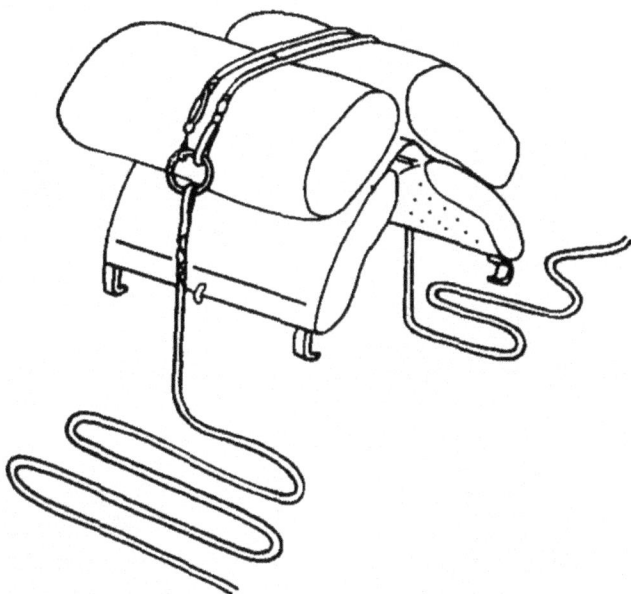

FIGURE 69. Load cinch positioned (Sweeten diamond hitch).

the running rope through the ring, outside to inside, and downward between the horizontal rope and the saddle (fig. 70). The running rope now is passed upward and to the outside of the horizontal standing rope, again through the ring, outside to inside, outside of the horizontal standing rope, and under the right footrest, left to right (fig. 71).

c. Each packer now passes his running rope upward over the top right corner of his side load, over and under his right rope of the load cinch part between the loads, and to the opposite packer. The hitch now is in position and ready to be tightened (fig. 72).

d. With both hands, left one·uppermost, the packers tighten the hitch as follows:

(1) Each grasps the inner of the two vertical ropes at a point just below the ring.

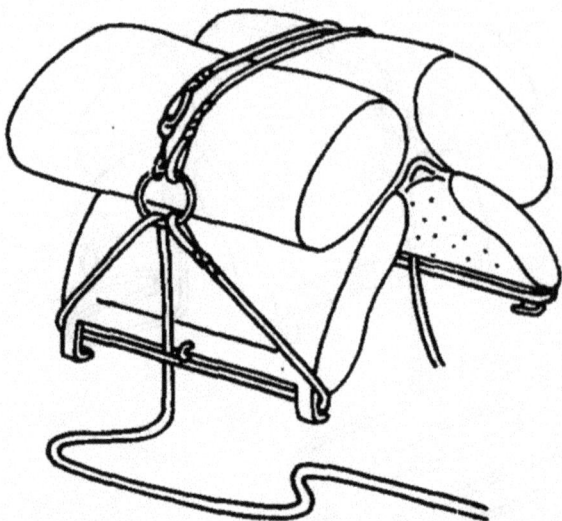

FIGURE 70. First step in forming Sweeten diamond hitch.

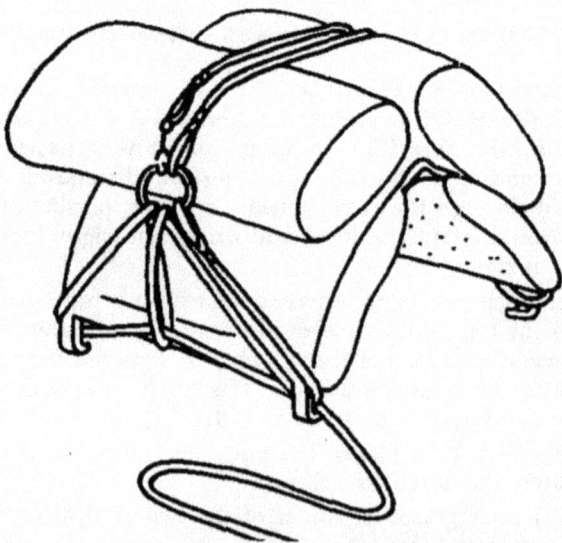

FIGURE 71. Second step in forming Sweeten diamond hitch.

96

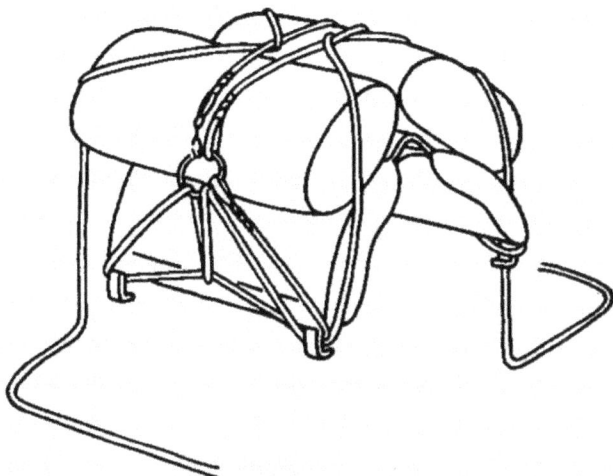

FIGURE 72. Sweeten diamond hitch in position ready to be tightened.

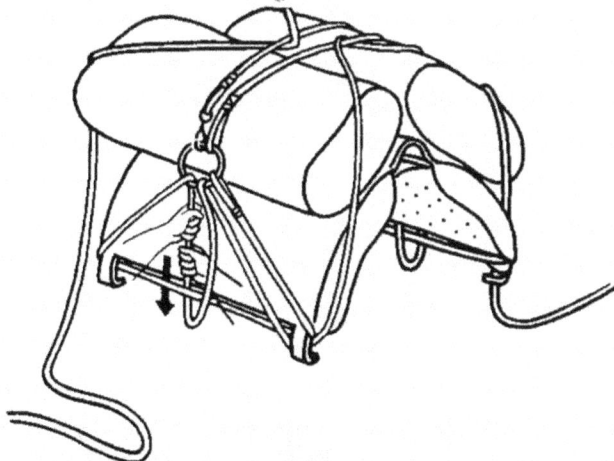

FIGURE 73. First step in tightening Sweeten diamond hitch.

(2) The near packer calls "Take slack." As soon as the off packer is ready, he calls "Down," at which time both packers pull strongly downward (fig. 73).

(3) Holding the tension with his left hand, each with his right hand, grasps the outer of the two vertical ropes at a point just above the horizontal rope, takes up slack, and then grasps it with both hands. The near packer calls "Take slack." The off packer, when ready, calls "Up," and each packer pulls strongly upward toward the ring.

(4) Each packer, holding the tension with his left hand, grasps with his right hand the right quarter rope just above the right footrest. Taking up the slack, he slides his right hand up the quarter rope and assures that it is correctly positioned over the end of his side load.

(5) Still holding the tension on the right quarter rope with his right hand, each packer grasps with his left

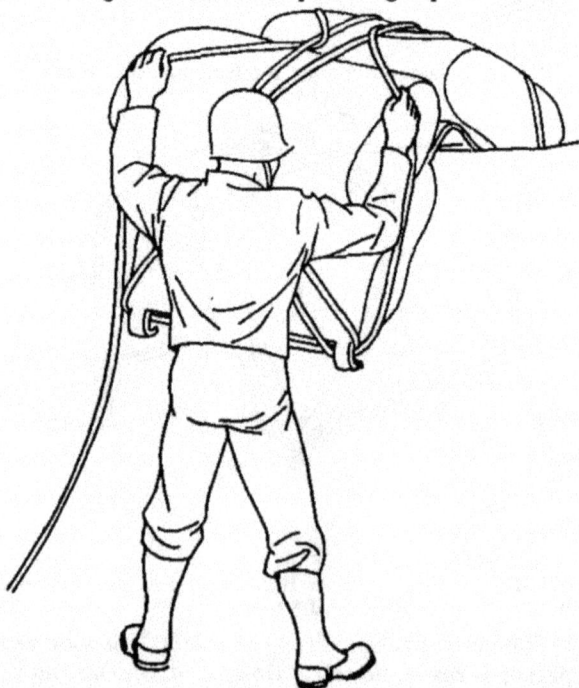

FIGURE 74. Holding tension on right quarter rope while taking slack on the running rope (Sweeten diamond hitch).

hand the running rope, which will form his left quarter rope, and takes up the slack from his partner (fig. 74).

(6) Releasing the right quarter rope as soon as he feels the slack being taken out of his rope by his partner, he sidesteps to the left and, facing to the right, places his left foot against his left footrest, and grasps the rope with both hands (fig. 75).

(7) The near packer calls "Take slack." When ready, the off packer calls "Pull," and both packers pull strongly on their running ropes. Each packer then repositions the running rope over the top left corner of his side load and forms his left quarter rope by passing the running rope under his left footrest, left to right.

(8) Each packer now hobbles his side of the hitch as described in paragraph 33i (figs 51A, B, and C), except that the hobble is tied with a slippery clove hitch around the right quarter rope.

e. Experienced packers frequently hasten the operation described in b above by rotating the spliced eye of the running rope through the 4-inch ring to form the necessary loops, instead of threading the end of the running rope through the ring. The ring being properly positioned on the load, the eye is rotated twice through the ring, *inside to outside,* to form the loops, keeping the eye splice to the right of the loops (fig. 76). The loop to the right is spread, right to left, through the left loop, keeping the eye splice to the right, and placed under the footrests (fig. 77).

40. NAGLE HITCH. The Nagle hitch, a modification of the Sweeten diamond hitch, is satisfactory for two side loads and can be formed more quickly than either the Phillips cargo or Sweeten diamond hitches. The lash rope for this hitch consists of two 30-foot pieces of ½-inch rope fastened to a 4-inch ring.

a. If the method of slinging described in paragraph 24 is used, the near packer gets the lash rope while the

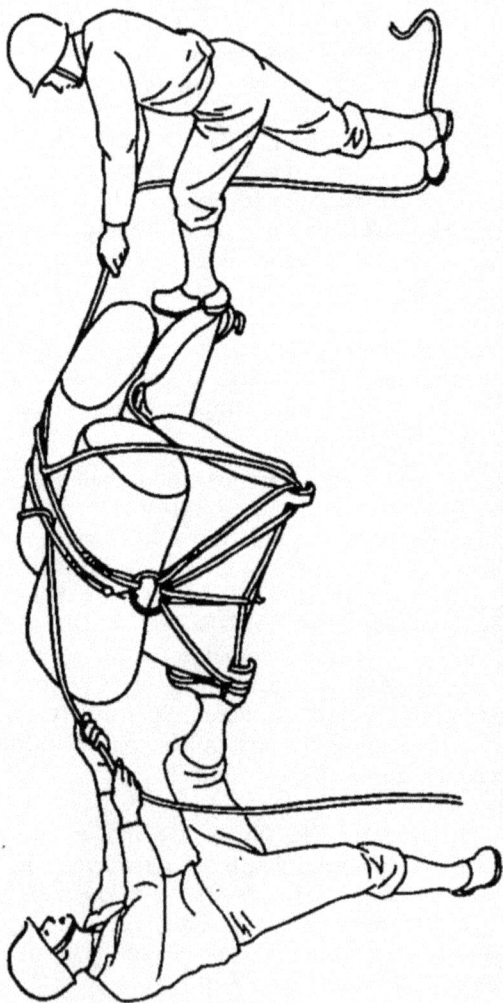

FIGURE 75. Both packers taking slack from their running ropes (Sweeten diamond hitch).

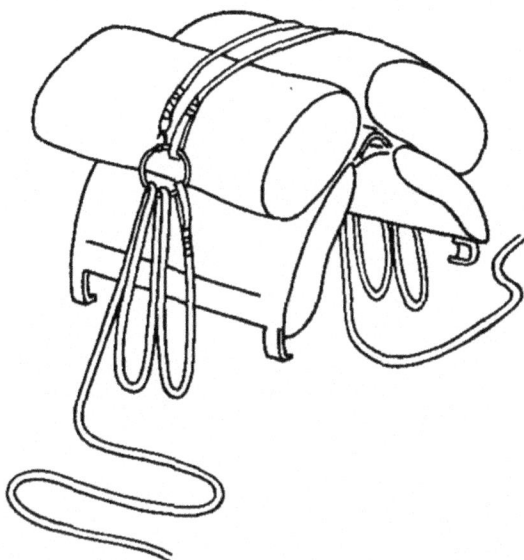

FIGURE 76. Loops for rapid formation of Sweeten diamond hitch.

off packer finishes tying the sling. While in rear of the animal, he checks the balance of the load.

b. Both packers stand facing each other in rear of the animal and both hold the ring horizontal with their left hands.

c. Having his eye splice and running rope to his right, each packer rotates his eye splice *once* through the ring from below upward, forming a loop which he then enlarges sufficiently to form the quarter ropes for his side of the load (figs. 78 and 79).

d. Each packer grasps his loop and running rope, with the running rope underneath, and both place the lash rope over the load, the 4-inch ring being centered between the two side loads.

e. Holding the ring in place with his left hand, each packer places his loop over the ends of his load and

101

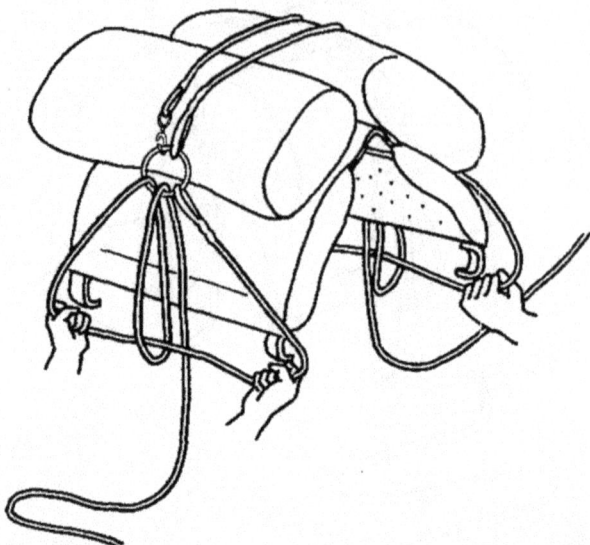

FIGURE 77. Second step in rapid formation of Sweeten diamond hitch.

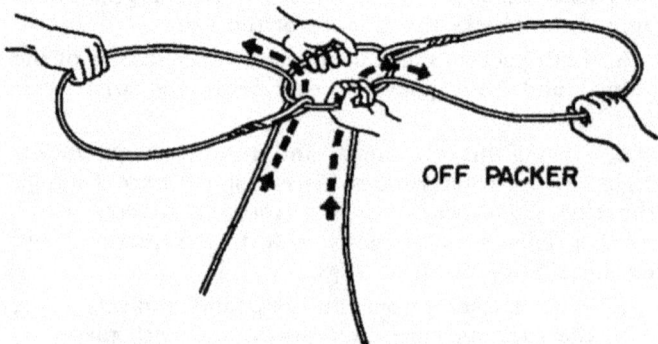

NEAR PACKER

OFF PACKER

FIGURE 78. Forming loops for Nagle hitch.

under the footrests, right to left, the free portion of the running rope dropping between the horizontal standing rope and the saddle (fig. 80). With both hands, each

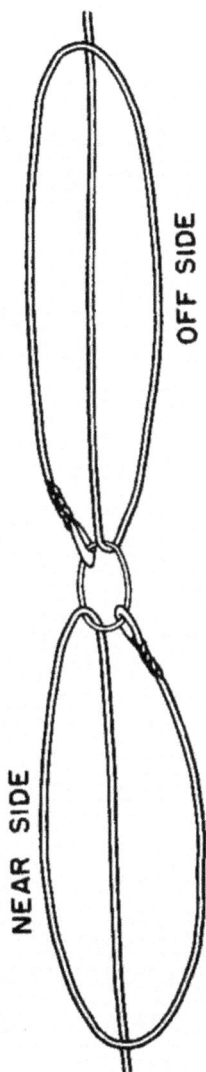

NEAR SIDE

OFF SIDE

FIGURE 79. Loops formed for Nagle hitch.

103

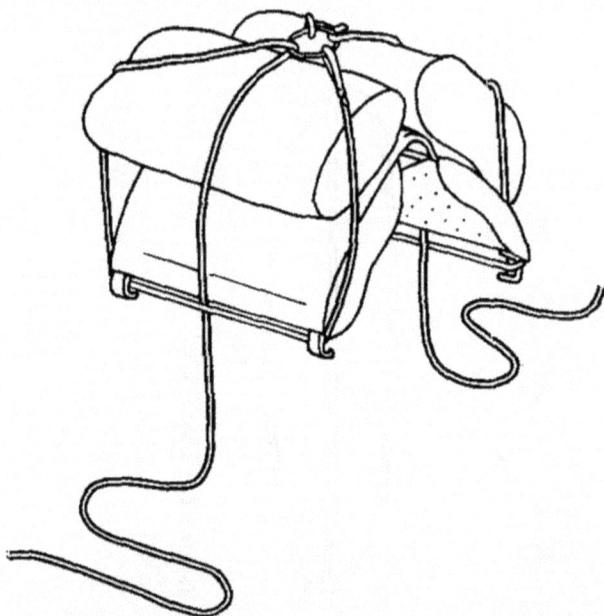

FIGURE 80. Loops in place on load (Nagle hitch).

packer now grasps his running rope just below the side load. The near packer calls "Ready." When ready, the off packer calls "Down," and both packers pull down strongly to seat the ropes.

f. With his right hand, each next grasps his running rope just below the horizontal rope, pulls upward, and holds the tension by binding the rope against itself with his left hand. He grasps the running rope with his right hand and passes it over the top of the load to his partner.

g. Each packer pulls the slack out of the running rope received from his partner. The near packer temporarily bights his rope, right to left, under the standing ropes where they cross the lower edge of his side load. The off packer, having received the running rope from his partner, does not bight the rope, but maintains tension.

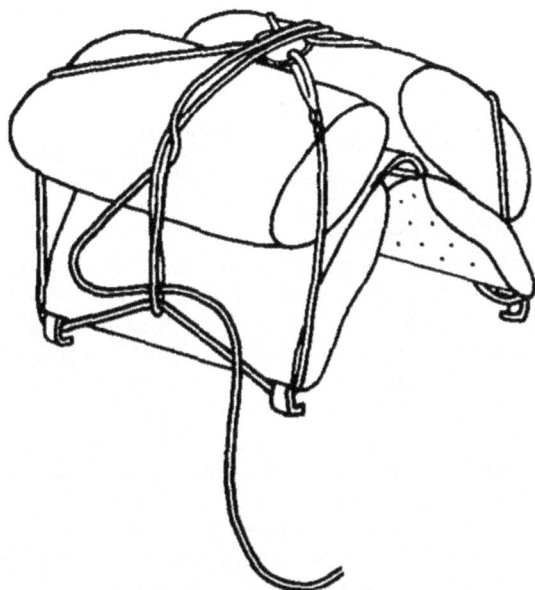

FIGURE 81. Nagle hitch in position and ready to be hobbled.

h. The near packer having completed the temporary bight, calls "Take slack," and the off packer calls "Go," whereupon the near packer tightens ("hits") the outer vertical standing rope at a point just above the horizontal standing rope.

i. The off packer, taking the slack, permanently bights his running rope under the standing ropes, right to left. He then calls "Take slack."

j. The near packer removes his temporary bight and, putting tension on the running rope, calls "Go." The off packer tightens ("hits") his outside vertical standing rope at a point just above the horizontal standing rope. The near packer pulls out the slack and permanently bights the running rope, right to left, under the vertical standing ropes (fig. 81).

k. Each packer then forms the hobble on his side as described in paragraph 32c (figs. 41, 42, 43).

CHAPTER 6
HANGER AND ADAPTER LOADS

41. HANGER LOADS.

a. Side hangers are equipped with hooks which fit over the hanger bars of the saddle. The hangers rest on the spring steel ribs and are tied down to the two end staples on the bottom bars. Hooks for hangers should fit loosely. All hangers and boxes should be equipped with distance pieces or rests riveted to the lower part of the box or hanger. The height of these pieces should be sufficient to clear the load from hooks to distance pieces. The load never should rest on the swell of the ribs. The "hanger distance," measured between the outside edges of the hooks, is 13⅝ inches for all Phillips pack saddles, cargo or cavalry.

b. In designing hangers for side loads of unequal weight, provision should be made for the heavy or bulky load to be placed where it will ride best, and the lighter or less bulky load fitted so as to obtain the desired balance. Where there is but a slight difference in the weight and bulk of the side loads, the load will balance if the heavier of the two side loads is placed slightly higher than the lighter one. Where a bulky side load extends considerably outward from the side of the saddle, although somewhat lighter than the opposite load, it generally will balance the opposite load. *Correct balance shoud be maintained at all times.* Heavy objects should be placed in the forward compartments of the boxes. If practicable, all loads should hang close to the saddle, and extend downward not much below the middle of the pads.

c. The arches of the saddle frame are designed to permit the attachment of top arch loads. There are two standard types of the arch: one for all heavy loads, such as the loads of pack artillery, and the other for light loads. Both types are bolted to the saddle through the two holes in each saddle arch. For all loads, a slight excess of the weight should be placed just forward of the center of the saddle.

42. ADAPTER LOADS

a. Special types of load arch occasionally are required for certain loads. Such arches are referred to as *adapters* and are bolted to the arches of the Phillips pack saddle.

b. In general, adapters are constructed to raise the load sufficiently high off the saddle to clear it from the animal's croup and neck, and to allow the load to be firmly fastened to the saddle. Adapter loads are secured to the saddle by load cinches or specially designed steel cables or clamps. The latter are integral parts of the adapter assemblies.

c. Adapter loads generally are high-riding top loads which require that the animals be led individually at the walk or amble.

CHAPTER 7
PACK TRAIN (HERDED)

43. GENERAL.

a. Herded pack trains, generally organized, trained, and operated by quartermaster and pack artillery units, are capable of moving large amounts of supplies. Loads usually are secured to the saddle by lash ropes.

b. Normally, the pack train operates as follows:

(1) The pack mules are herded. They are trained to follow in single file, without crowding, the sound of the bell carried by a led mare. The bell mare always should be led with reins and bit.

(2) The rate of march is 4½ to 5 miles per hour. Mules marched at this rate soon will acquire the ambling gait which is smoother than a walk and causes a minimum of rocking motion of the load.

(3) Once on the march, halts should be made only in an emergency. In such event, previously designated round-up men control the herd while needed adjustments are made. Upon resuming the march, a check is made to insure that no animals have strayed.

(4) Packers are distributed along the column to watch the loads. When an adjustment is necessary, the mule must be caught, led out of column, and the adjustment made by two packers working together. *Packs must be adjusted rapidly.*

(5) On trails, packers ride in the column distributed so that each can watch about five pack mules.

(6) In rough country, several packers ride far enough in advance of the head of the column to enable them to station themselves at dangerous places. This is a precaution taken to keep the mules on the trail and

prevent accidents. The mule loaded with the pioneer tools should be led at the head of the train.

(7) Watering herded pack animals on the march is difficult. If time permits, the column should be halted and animals led to water. If it is impracticable to do this, care must be exercised to insure that the mules do not lie down in the water and damage their loads.

44. DUTIES OF INDIVIDUALS.

a. Packer. Pack train personnel should be selected from men who like animals and are accustomed to hard work. They must be trained to perform their duties rapidly and skillfully. The packer is an understudy of the cargador and should be able to perform the duties of the cargador in the latter's absence. The packer's specific duties are to—

(1) Train and saddle pack animals.
(2) Train, saddle, and ride riding animals.
(3) Care for animals in the field.
(4) Properly care for and use saddle equipment.
(5) Prepare cargo into loads for packing.
(6) Form all hitches used in packing.
(7) Sling and lash loads.
(8) Remove loads.
(9) Tie and splice ropes and cord.

b. Cargador. The cargador assists the packmaster in all his duties and should be able to perform these duties in the latter's absence. In addition, he may be the saddler and, as such, is responsible for all repairs normally made by the saddler. For additional information on the duties of the saddler see TM 10–226 and 10–430. The cargador's specific duties are to—

(1) Assign pack mules and equipment to the packers.
(2) Instruct packers as to the type of load for each pack mule.
(3) Match up cargo to make balanced pay loads.
(4) Maintain strict order and discipline among the packers.

(5) Require quiet and gentle treatment of pack mules.

(6) Select areas for cargo piles in bivouac.

(7) Assist the packmaster in working saddle pads.

(8) Keep a memorandum of all cargo and equipment under his care, marking and tagging it if necessary.

(9) Insure that all pack equipment is properly cared for in bivouac.

(10) Select areas for saddles in bivouac and remove saddles as the animals are brought to him.

(11) Pay off (untie and release) mules from the floating picket line and assign saddles and loads as he does so.

c. Duties of packmaster. The packmaster is responsible for the presence, care, and maintenance of all pack equipment and animals of his unit. He rides the entire column in order to check all loads and observe the condition of men and animals. In some organizations, he also performs the duties of stable sergeant. The packmaster's specific duties are to—

(1) Exercise general supervision over all packing and pack loads and require that loads be properly packed to avoid injury to the animals' backs.

(2) Train personnel under him in the proper method of saddling, adjusting, and packing the pack saddle.

(3) Require proper care of pack animals at all times.

(4) Select and assign pack and riding mules.

(5) Fit all pack saddles.

(6) Require readjustment of loads whenever necessary.

(7) Check pack animals for injuries when pack saddles are removed.

(8) Work and adjust pack saddle pads to maintain fit and to relieve injuries.

(9) Require all pack equipment to be kept in good condition.

(10) Insure that all breakage or damage to the pack saddle is repaired.

(11) Check animals on the march for signs of distress or weakness and, if necessary, relieve them of loads.

(12) Assist the train commander in organizing personnel and animals at stream crossings.

d. Train commander. The train commander is the officer placed in charge of the train. His specific duties are to—

(1) Assume responsibility for the conduct of the train.

(2) Train and discipline personnel and assign them appropriate duties.

(3) Enforce strict care and conditioning of animals.

(4) Require proper care and maintenance of all pack equipment and cargo under his control.

(5) Exercise close supervision on the march.

(6) Enforce measures for proper cover, concealment, and protection against surprise attacks by enemy air or ground forces.

(7) Obtain information as to the location of the forward echelon.

45. STABLES AND CORRAL FOR PACK TRAIN.

a. In garrison.

(1) In addition to the day corral, a work corral, equipped with saddle and cargo racks, should be provided. The top of the pack saddle rack (rigging rack) should be from 6 to 12 inches above the ground level. Each cargo rack should be large enough to hold about 20 pack animal loads. A corral of the recommended type is shown in figure 82.

(2) In warm weather, feed racks should be provided in the day corral.

(3) In cold weather, the animals should be fed inside.

(4) Stables should provide—

(a) A box stall for the bell mare.

(b) Mangers with tight bottoms for pack mules. If possible, the mangers should be edged with metal.

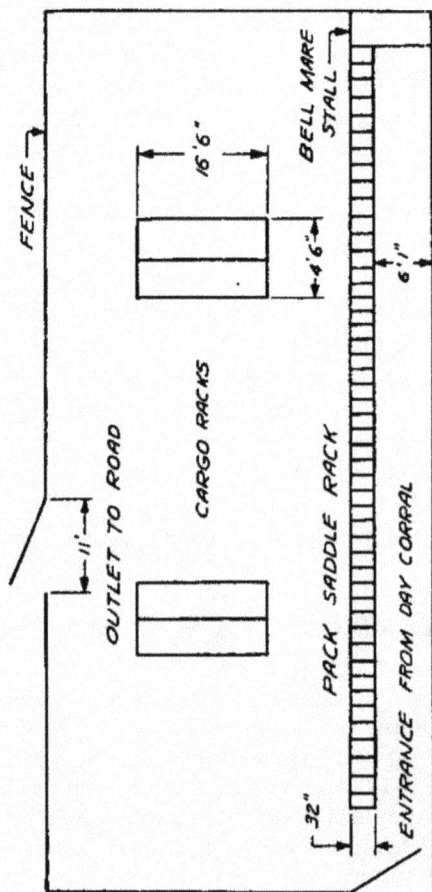

FIGURE 82. Suggested plan for pack train work corral.

(c) Doors on the corral side of the stable. They should be fastened open to allow animals free access to the stable.

(d) Padding up to 5 feet from the floor on all posts. Mangers and sharp corners should be rounded off and padded.

112

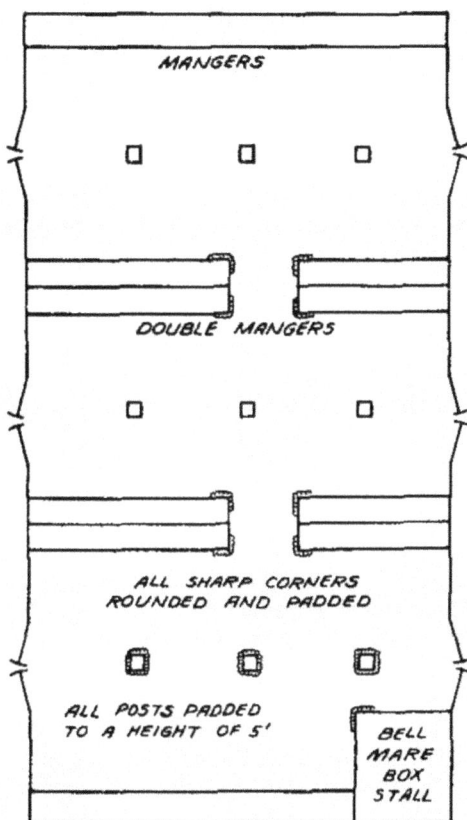

FIGURE 83. Suggested plan for stable for herded animals.

(5) A pack train stable of the recommended type is shown in figure 83.

b. In the field. When a pack unit occupies a semi-permanent camp, it should build—

(1) A pole corral.

(2) Mangers and feed racks, if material is available. If not, animals, tied together on a floating picket line, should be fed grain on rigging covers.

46. LINING UP PACK MULES. Herded mules are lined up primarily for the purpose of making it easy to catch them. If properly trained, mules will line up rapidly and without confusion.

a. In garrison. All pack mules, including those not to be worked, should be lined up each day. Lining up is accomplished by steps as follows (figs. 84, 85, 86, 87):

(1) Groom and saddle riding mules and tie them on the picket line.

(2) Remove the cargo and rigging covers.

(3) Place halter shanks on the pack saddles, snaps toward the animals to be lined up later.

(4) Lead the bell mare to the entrance of the chute in rear of the pack saddles and hold her there until all mules have been closed to that end of the corral.

(5) Two packers hold a lash rope across the end of the corral to prevent the mules from running back and to force them into the chute. The rope should be kept in motion up and down (fig. 84).

(6) Two packers stand in rear of the fence and cause the mules to close properly. One packer sees to it that the mules face the saddles; the other urges them toward the bell mare.

(7) As the bell mare nears the end of the line, she is turned temporarily to face the saddles in order to check the push of the on-coming mules (fig. 85). She then is eased over into her stall, secured, and bridled.

(8) Snap halter shanks to the halters and tie animals together to form a floating picket line (par. 47).

(9) *Stress quietness in lining up the mules.*

b. In the field. The garrison procedure is modified in the field by (fig. 88):

(1) Improvising the fence in rear of the saddles by lash ropes tied to trees or held by packers.

(2) Having mounted drivers herd the animals into the chute formed by the lash ropes and the line of saddles.

(3) Using a tree on which to tie the bell mare.

(4) Lining up animals so they do not face into high wind, rain, or sleet (figs. 89, 90, 91, 92).

FIGURE 84. Mules following bell mare into chute.

FIGURE 85. Bell mare being turned to face saddles.

FIGURE 86. Mules being closed toward bell mare and turned to face rigging.

FIGURE 87. Forming the floating picket line.

FIGURE 88. Using lash ropes to form field chute.

119

FIGURE 89. Lining up mules in field chute.

FIGURE 90. Turning mules to face rigging in field chute.

FIGURE 91. Forming the floating picket line in field chute.

122

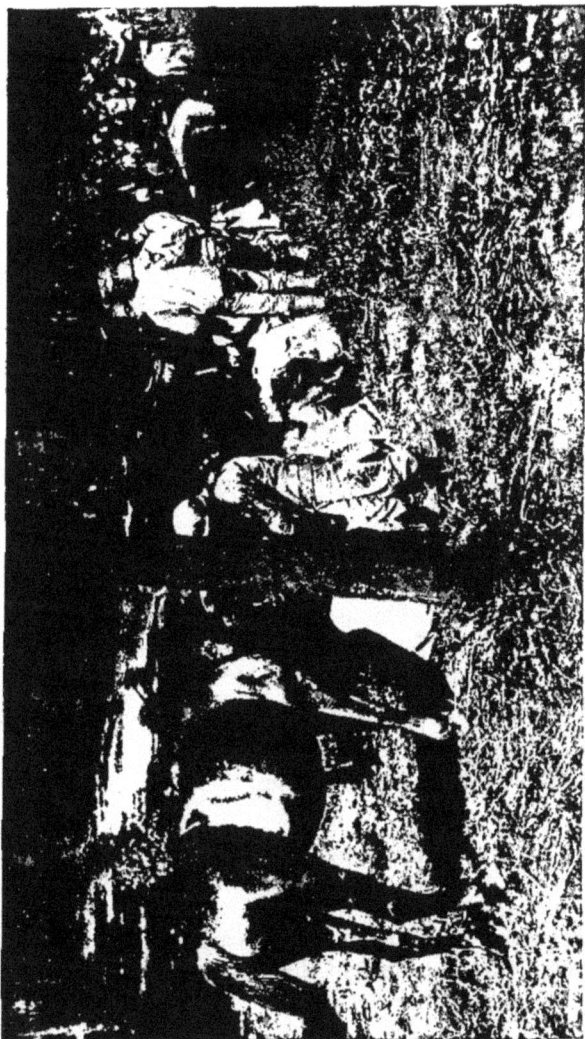

FIGURE 92. Feeding grain in the field.

123

c. Remount mules or mules new to the train. The following procedure is recommended:

(1) Herd the mules in rear of the bell mare through the chute several times to accustom them to the chute.

(2) Leave the rigging covers on the saddles at first and place a few oats thereon to encourage the animals to turn toward the saddles.

(3) Require men to form in line approximately 3 yards in front of the saddles to prevent the mules from jumping over them.

(4) Station one man with a long switch in rear of the line of mules to cause mules to close *gently* and *quietly* to the right and turn toward the saddles.

(5) Tie the leading mule to the bell mare's stall and then successively line up each mule.

(6) Packers on the line step forward *quietly*, snap the halter shanks to the halters, and tie the shanks together to form a floating picket line.

47. FLOATING PICKET LINE. When pack animals are lined up to the saddles (rigging) in the preparation for work or feeding, they are tied together to form a floating picket line. Packers accomplish this as follows:

a. Snap halter shanks to the halter tie rings.

b. Form a 6-inch loop approximately 20 inches below the snap on the halter shank.

c. Hold the loop in the left hand.

d. With the right hand, grasp the halter shank of the mule on the right at a point about 4 feet from the snap. Lay the rope over the loop held in the left hand, bend it under the loop and, forming a loop with the running end, pass it over the standing part, left to right, and through the loop held in the left hand. Draw the knot tight (fig. 93).

e. Proceed in foregoing manner until all mules are tied together.

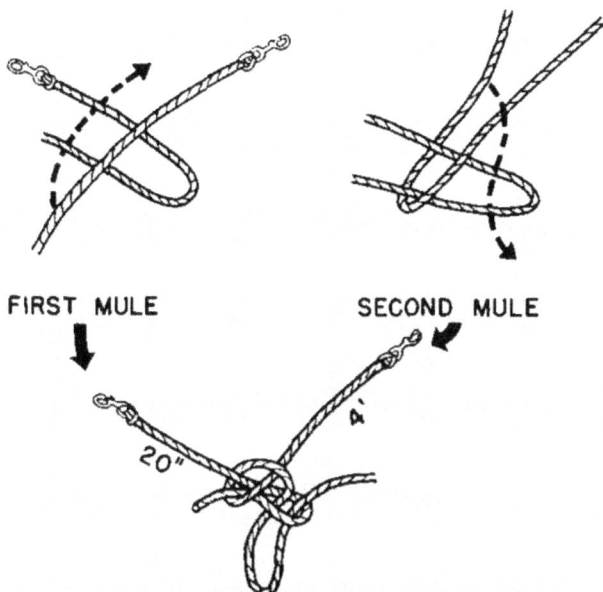

FIRST MULE   SECOND MULE

FIGURE 93. Floating picket line knot.

48. SADDLING AND PACKING IN PACK TRAIN.

Mules habitually are packed as soon as they are saddled. Working in pairs, packers proceed as follows:

a. To each near side packer, the cargador pays off a pack mule, giving him the work number and any special instructions concerning the load. The work number assigned to identify both the mule and the equipment, is painted on the saddle cover and halter of the mule for which the saddle is fitted. A record is kept of the work number and the Preston brand number of each mule.

b. The near packer, assisted by his partner, leads the mule and places him so that he faces the saddle directly to the right of his own saddle.

c. The off packer puts the woven pad in place. If necessary, the near packer applies the blind to the mule.

d. The near packer lifts the saddle off the rigging rack and, grasping the near hanger bar and a near foot-rest, tilts the saddle away from him so that the near bottom bar rests on his thigh and the off footrests are on the ground (fig. 94). He then swings the saddle up and over the mule's back, being careful not to strike the mule. The off packer catches the off footrests. Together, they gently lower the saddle into place (fig. 95).

e. Packers proceed as described in paragraph 14.

f. The blind, if used, is raised and the animal led to the cargo and made to face the side of the pile. If necessary, the blind is replaced.

g. When the hitch is finished, the off packer receives the blind from the near packer, obtains another mule, and becomes the near packer for the next load. Meanwhile, the near packer ties up the halter shank, releases the mule, and becomes the off packer for the next load.

49. UNPACKING AND UNSADDLING IN PACK TRAIN.

a. All mules are unpacked before a single one is unsaddled. Following the removal of the load from the saddle, the cinches should be loosened slightly and the lash and sling ropes coiled.

b. When all loads have been removed, the packers catch the mules and prepare them for the removal of the saddle by untying the thongs on the woven pad, releasing the cinches and securing them over the top of the saddle, and unsnapping the holding straps and turning the breeching over the top of the saddle.

c. The mules now are led (in numerical order) to the cargador who removes the saddles and pads from the off side and places them in their proper places in line. In removing the saddle, he is assisted by the packer who lifts upward on the near footrests (figs. 96 and 97). The mule is then either released in the corral (or herd) or tied on the picket line.

FIGURE 94. Near packer preparing to lift saddle.

FIGURE 95. Near packer lifting saddle prior to placing it on animal's back.

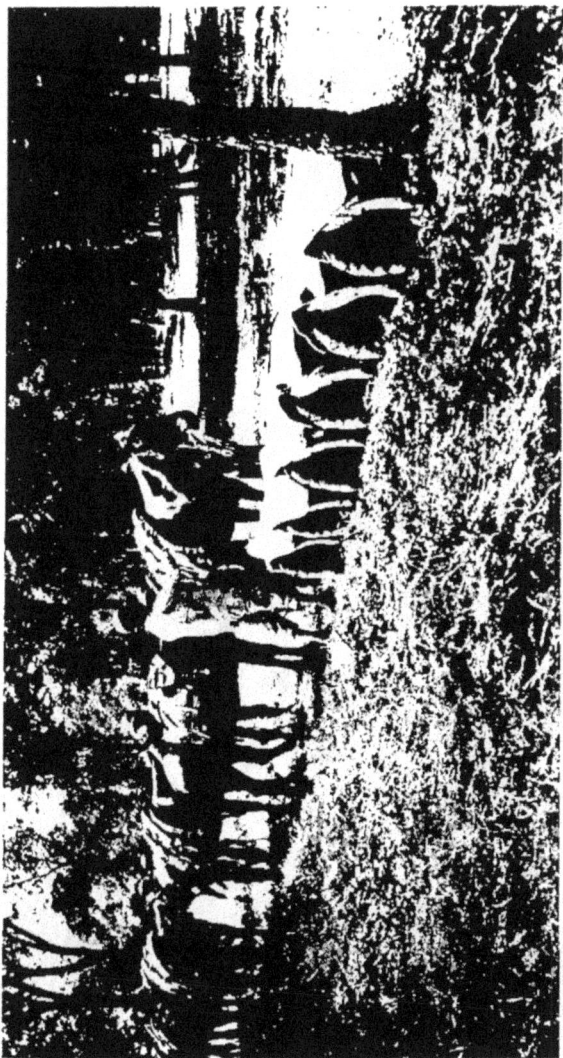

FIGURE 96. Mules being led to cargador for unsaddling.

FIGURE 97. *Cargador unsaddling mule.*

50. TYING UP HALTER SHANKS.

a. Pack mules. The pack mule's halter shank must be tied up without slack so that he will not become entangled in it. This is accomplished in the following manner:

(1) With the snap of the halter shank attached to the halter tie-ring, pass the halter shank over the animal's neck to the off side.

(2) Form a small loop in the halter shank on the near side and pass it, inside to outside, through the near cheek ring of the halter.

(3) Reach under the animal's neck and bring the loose end of the halter shank up under his throat.

(4) Form a loop in this part of the shank and pass it, right to left, through the loop already inserted in the cheek ring. *The rope around the neck should fit snugly.*

(5) Continue forming loops as in (4) above until only a short free end remains to be pulled through the last loop (fig. 98).

b. Riding mules. The riding mule's halter shank is tied so as to enable the rider to use it in leading a pack mule. It should be tied as follows:

(1) Attach the snap, inside to outside, to the near cheek ring of the halter.

(2) Allowing only sufficient slack to enable the mule to lower his head to graze, fasten the free end of the shank to the saddle horn with a clove hitch.

(3) To lead a pack mule, the rider unsnaps the shank from his mule and snaps it to the halter tie-ring of the pack mule. He then takes the top hitch from the saddle horn and holds the loose end of the halter shank in his right hand.

51. MARCHES.

a. Departure from bivouac. Upon completion of packing, all men mount their riding mules and take positions as follows:

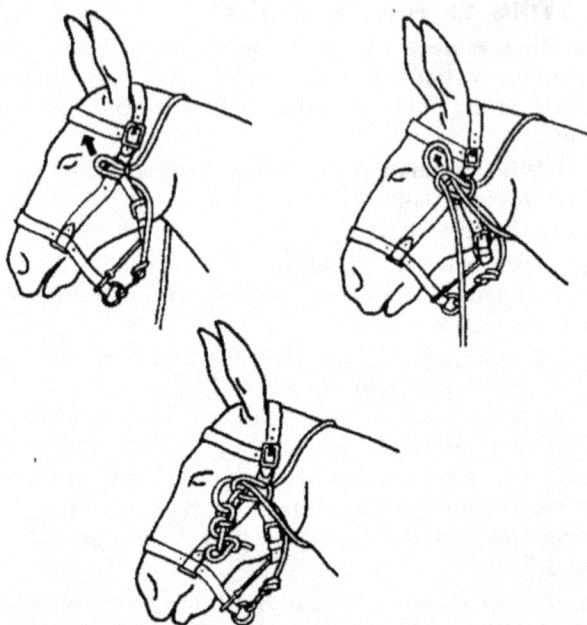

FIGURE 98. Tying up the halter shank.

(1) The bell mare driver leads out as directed.

(2) Two packers ride ahead of the train as test riders and guards.

(3) For road and wide trail marching, two packers place themselves behind each ten mules.

(4) For narrow trail marching, one packer places himself behind each five pack mules.

b. Conduct of the march.

(1) *General.* While on the march, the mules are herded behind the bell mare. Normally, they string out in single file; however, they should not be allowed to string out at such distances that the packers preceding and following each ten mules cannot see or hear each other. In general, rules to be observed in controlling the herded march are as follows:

(*a*) When passing other herded pack trains, all packers ride on the side of the column nearest the other train.

(*b*) All personnel should be required to remain close to the train, leaving the column only to adjust a load, or upon orders of the train commander.

(*c*) Mounted personnel never should be allowed to approach the halted train at a gait faster than the trot.

(*d*) While packing or at the halt, smoking and loud noises should be prohibited.

(*e*) Abuse or mistreatment of animals must not be tolerated:

(*f*) Strange animals, regardless of sex, should not be allowed to come near the train.

(*g*) Cattle, burros, and small animals such as hogs, sheep, goats, and dogs should be kept at a distance on the march and not allowed in camp.

(2) *At night.* In addition to the general rules in (1) above, the following rules are applicable when marching at night:

(*a*) It is extremely important that the column be kept well closed up to facilitate control of the train and to make it possible to count the mules frequently.

(*b*) Straying is prevented by riders at the flank and rear of the train.

(*c*) It generally is better to reduce the rate of march.

(*d*) Hourly or more frequent halts, ordinarily not made in the daytime, are made to inspect and adjust loads and saddles. Inspection of loads and saddles must include feeling by hand for adjustment.

c. Supervision by officers and noncommissioned officers. The train commander, packmaster, and cargadors should alternate in making trips to the head of the train for inspection of it as it passes. This inspection should include a count of the pack mules, a check of the condition of mules and loads, and the correction of any faults noted.

d. Packers. In addition to their other duties incident to the maintenance of march discipline, packers are responsible specifically for the proper adjustment of loads and equipment. Pack mules are caught for the adjustment of loads as follows:

(1) Two packers proceed 50 to 100 yards (preferably to a narrow place in the trail) ahead of the pack mule whose load needs adjustment.

(2) They dismount, secure their riding mules, and one passes through the train to the opposite side.

(3) As the pack mule approaches, both packers close in *quietly*, catch the mule, and lead it to the side of the trail on which the riding mules were secured.

(4) The load or saddle is adjusted as rapidly as possible.

(5) The adjustment having been completed, the packers mount their riding mules and lead the pack mule to the rear of the column at a gait no faster than the trot.

e. Special duties. When on the march, packers perform various special duties which are rotated as far as practicable from day to day as follows:

(1) *Bell mare driver.* His duties include—

(a) Leading the bell mare at the head of the train.

(b) Silencing the bell when passing other pack trains or near the enemy.

(c) Keeping the bell mare quiet and in place while train is halted.

(d) Tying up the bell mare upon entering the corral or bivouac.

(2) *Roundup men.* Their duties include—

(a) Packing and unpacking except when the herd is at a march halt.

(b) Assisting with adjustments of loads while en route.

(c) Remaining mounted at the halt to circle and round up all animals that stray from the train.

52. BIVOUACKING PACK TRAIN.

a. The train commander precedes the train into bivouac and selects areas for the rigging line, the picket line, and the cargo. Upon arrival of the train, the routine is as follows:

(1) All packers, except the roundup men, dismount, tie up their riding mules, unload the cargo, and "slack off" cinches.

(2) As they are unloaded, mules with lower work numbers are tied near the area designated for the line of pack saddles (rigging). The other mules are tied near the opposite end of the line.

(3) When all loads have been removed, the train commander gives the command to unsaddle.

(4) If the soil is suitable, all pack and saddle mules are released so that they may roll, thereby resting and massaging their back muscles.

(5) Time available prior to watering is used for cleaning equipment or in preparing the camp.

(6) Animals are watered as soon as they have cooled off, normally about 45 minutes after arrival at bivouac.

(7) After watering, all animals are groomed thoroughly.

(8) When the animals have been cared for properly, all broken or damaged equipment is repaired and made ready for the next day's march.

b. Mules, equipment, and cargo of the pack train are arranged systematically so as to be readily available, night or day. If the situation will permit, saddles are arranged as in the garrison work corral. Cargo is located similarly; riding saddles may be placed on top of the cargo. Saddle and cargo racks need not be provided unless the unit occupies a semipermanent bivouac.

c. When adequate cover is not available, or greater dispersion is desired, the train may be broken down into smaller units, each arranging its own mules, equipment, and cargo.

53. TRIPS TO DISTRIBUTING POINT.

a. Preparation of train. When traveling to the DP without cargo, items, such as cargo covers and ropes, are wrapped and packed as single loads. Each load should contain the equipment for five pack mules.

b. Procedure at distributing point.

(1) All mules are caught and tied.

(2) All cargo to be wrapped in cargo covers is wrapped before any is packed.

(3) The cargo then is matched into balanced loads by the cargador.

c. Delivery of cargo. Cargo is delivered to the unit or units as ordered.

CHAPTER 8
PACK TRANSPORT WITH INDI-
VIDUALLY LED ANIMALS

54. GENERAL.

a. In pack artillery, infantry, engineers, and certain other units, pack animals usually are led by dismounted drivers. These animals may be employed to transport equipment, supplies, and cargo of all kinds; however, they generally are employed to carry high-riding top loads secured to the saddles by load arches, adapters, or ropes.

b. The rate of march is 3½ to 4 miles per hour, which allows the mules to take their natural gait.

c. The progressive conditioning and training of both men and animals are vitally important.

55. DUTIES OF INDIVIDUALS. The duties of pack-
masters, cargadors, and packers are similar to those described for the pack train (ch. 7). The drivers should be trained to pack lashed loads as well as hanger- and adapter-positioned loads.

56. LEADING ANIMALS ON THE MARCH.

a. When pack animals are led, they should be allowed freedom of movement and balance and, at the same time, controlled sufficiently to keep them in their proper places in column. Animals usually are led from the near side; however, when the column is marching so as to expose them to noise, danger, or confusion on the off side, they should be led from the off side.

b. For leading animals over flat and even terrain, packers grasp the reins in their right hands, 6 inches from the bits, and hold the remainder in their left. This is reversed when leading from the off side. Drivers must not let the animals pull them along. Animals are led at a steady pace and *never* allowed to walk and trot alternately. The prescribed distance must be maintained constantly; *accordion action in the column causes fatigue in the rear elements.*

c. When animals are led through small ditches or ravines, or down short slopes, their heads are held down to keep them from trotting; however, the rate of march must be maintained. The mules should not be allowed to jump over such obstacles as ditches, logs, and boulders. Jumping causes displacement of the load.

d. When animals are led up steep slopes or over very rugged terrain, they are given their heads as much as possible so as to allow them to seek their own footing and maintain their balance. In all such cases, the driver must stay far enough ahead to keep out of the animal's way. An allowance of 1 yard of loose rein is the normal minimum. If the terrain is very rough or steep and the driver should fall behind, it is better to drop the reins and let the animal go; he is caught as soon as the obstacle has been passed.

e. Packers and other personnel not leading animals, march near the flank of the animal to which each is assigned. They watch the adjustment of the load and the saddle, and help keep the animal closed up to his proper place in column. *Under no circumstances will personnel hold on to the saddle, breeching, or animal's tail to assist themselves in walking.*

CHAPTER 9
MARCHES AND BIVOUACS

57. GENERAL.

a. During training, every effort should be made to condition both men and animals on the type or types of terrain over which they are expected to operate. All men should have a thorough knowledge of how to take care of themselves and their animals, both on the march and in bivouac. During training, marches initially should be short, and then gradually lengthened as the condition of both men and animals improves. Conditioned units must continually maintain marching schedules, at least 3 or 4 marching days per week, in order to be in constant readiness for extensive field service.

b. The tactical situation may make a certain rate, formation, or timing of march necessary. In the absence of restrictions and with road space available, a march should be conducted in such a formation and at such a rate that it will cause minimum fatigue to men and animals. However, it is important to complete the march and to relieve the animals of their packs at the earliest practicable moment. *Excessive rates of march should not be used unless specifically required by the tactical situation.*

c. All personnel should be furnished information concerning the destination, route, distance, rate of march, scheduled halts, and route markings.

d. For definitions of *distance, road space, time length, rate, gait of march,* and *march unit,* see TM 20–205.

58. RECONNAISSANCE OF ROUTES.

a. A route reconnaissance is made prior to a march over unknown ground. This duty should be performed by an officer, with sufficient men and pioneer equipment for making necessary minor repairs to trails, bridges, and other obstacles. If it is impracticable to reconnoiter the entire route, it is advisable to cover at least one day's march ahead of the organization.

b. Every effort should be made to push reconnaissance sufficiently far to the front to insure—

(1) Selection of passable routes.

(2) Indication of all necessary pioneer work. When the reconnaissance party finds an obstacle which cannot be bypassed by the column, an estimate of the labor, materials, and time required to make the route passable must be sent to the column commander.

(3) Length of the route being held to the minimum.

59. PIONEER WORK.

a. Pack units normally perform considerable pioneer work along routes over which they operate. It is important that training include exercises involving pioneer work of varied types.

b. If the route of march must cross marshes or swampy ground, time is allowed for the necessary pioneer work. One way to cross boggy ground is to make a trail of mats of brush, small trees, or logs. Care must be taken to insure the matting being sufficiently thick to support animals with loads. After a number of animals have crossed, such matting has a tendency to fall apart and break through; therefore, it is essential to observe it continuously, strengthening it periodically as necessary. Over exceptionally swampy ground, it is advisable to construct a low bridge, 18 to 24 inches wide. Such a bridge may be built by placing two logs, side by side, on piling, binding them together to prevent spreading, and filling in the crevice with brush and dirt.

c. Cutting trails through undergrowth and jungles may be necessary. In working such a trail, the pioneer party should be relieved frequently by fresh men so as to maintain a steady forward progress. The trail should be cut sufficiently wide and high to give clearance for all loads.

d. It may be necessary to construct trails over mountainous terrain. Blasting equipment is a valuable aid. On steep slopes, the trail should be cut so as to keep the degree of incline to a practical minimum. The following are specific suggestions:

(1) Construct switchbacks and landings on the trail in a zigzag formation, or run the trail on a long, upward traverse across the face of the slope.

(2) Where turns or switchbacks in the trail are adjacent to dangerous precipices, guard rails should be constructed of logs or rocks.

(3) Extreme caution should be observed on very steep slopes, especially if snow or ice is present, to prevent landslides.

60. MARCH DISCIPLINE.

a. General. An organization with good march discipline passes over routes with a maximum of speed and comfort and with a minimum of interference with other troops. It arrives at its destination with its personnel, animals, and materiel in the best condition permitted by the situation. Straggling, falling out of column, lounging in the saddle, and failure to keep to the proper side of the road are evidences of poor march discipline.

b. While marching. The principal duties of officers and noncommissioned officers while on the march are to insure that—

(1) Restrictions on the use of roads, lights, and radio are obeyed.

(2) Maximum use is made of cover and concealment.

(3) A uniform gait is set at the head of the column.

(4) Loads and saddles remain properly adjusted. *Most galls, abrasions, and bunches or bruises can be eliminated by careful and active supervision by all officers and noncommissioned officers while on the march.*

(5) No animal is overworked or overheated. A clinical therometer often is of the greatest service. *The animal's temperature is the surest index of his fitness to continue work.* The temperature of a horse or mule under normal conditions is from 98.5° to 101.3° F. It rises with exertion. If this rise is small, there is no danger, but if the temperature reaches 103° to 104°, precautions must be taken. If there is a further rise to 105°. the animal should not be worked.

(6) Distances are preserved so far as the terrain will permit.

(7) Drivers and animals maintain the prescribed gait and rate of march.

(8) There is no buckling in the column and that pack animals do not walk and trot alternately.

(9) *The rate of march is not changed when negotiable obstacles or inequalities in the trail are encountered.*

(10) Animals which fall out of the column for any reason clear the road so as not to interfere with other troops.

(11) The column keeps to the proper side of the road, leaving half the road clear for other traffic.

(12) No men straggle, mounted men do not lounge in the saddle, and that no one leaves the column without authority.

(13) Men do not obtain water from unauthorized sources.

(14) During hot weather, men do not drink excessive amounts of water while marching.

c. At the halt.

(1) Halts should be made at places which provide maximum cover and concealment. Local security is established.

(2) The time and length of halts should be made known in advance in order that full advantage may be taken of the opportunity to adjust packs and rigging and care for the animals. During the halt, drivers sponge

out the animals' nostrils, rub the legs, and clean out the feet, while the packers adjust the saddles and loads.

(3) A "shake-down" halt should be made between 15 and 30 minutes after starting a march. Sufficient time should be allowed at this halt to permit every load to be carefully inspected, adjusted and, where necessary, removed and relashed. The time spent is not lost since the precautions taken will prevent trouble later in the march.

(4) Thereafter, halts of 5 to 10 minutes should be made each hour. (These halts normally are not made by herded units.)

(5) Care should be taken in selecting places for halts, particularly the first halt of the day, in order that men may relieve themselves.

(6) Halts, unless of sufficient duration to allow all loads to be removed, give little relief to the pack animals. On long marches, either as to distance or time, a midway halt of at least one hour should be made in order that loads may be removed and men and animals rested and fed. Pack saddles should not be removed but the cinches should be moderately loosened at this halt. Six hours is considered a maximum length of march on which no midway halt is needed.

61. STREAM CROSSING.

a. Pack units frequently have to cross streams or bodies of water where no bridges exist. Stream crossing methods are vitally important and should be practiced in training.

b. When herding, the bell mare should be taken across first to encourage timid animals either to swim or ford the stream.

c. (1) Mules and horses can ford fairly deep streams; however, great care should be exercised in fording with loads, because the load makes the animal top-heavy and, if he loses his footing, he may turn over on his side and drown. Reconnaissance of fords and improvement of footing by advance details are necessary. If the current is swift and the water deep enough to bring pressure against the body of the animal, or if the footing is poor, the loads and saddles should be removed prior to fording.

(2) When fording, men should be posted on the downstream side to prevent animals from getting into dangerous places. Lash ropes, stretched across the stream on either side of the ford, also will assist in keeping animals on the proper course.

d. Although animals generally are good swimmers, they should not be swum while loaded or saddled. Loads make them top-heavy and they may become entangled in the cinches or breeching. In addition, when submerged, the saddle soaks up large amounts of water, making it very heavy.

e. Equipment, saddles, and cargo of all types may be ferried by boats or rafts across unfordable bodies of water. A serviceable boat may be made from unit equipment as follows:

(1) Unload all saddles and cargo as near the water's edge as possible. (At this time, all animals should be swum across.)

(2) Spread a rigging cover on the ground near the water. Place five pack saddles, end to end, with their footrests on the edge of the cover, and centered so that there will be an equal length of cover at either end. The saddles are lashed firmly together by a lair rope looped through their arches (fig. 99). The saddles then are rolled over as one unit to the center of the canvas.

(3) The footrests now are tied together with a lair rope, looping together all adjacent footrests (fig. 100).

(4) The rigging cover is lifted until it fits snugly around the outside of the saddle, the ends being folded and placed across the end saddles.

(5) The cover is bunched at each footrest and secured in place with a loop in the lair rope (fig. 101). This

144

FIGURE 99. Rigging cover spread and saddles tied arch to arch.

FIGURE 100. Saddles tied footrest to footrest ready for covering with rigging cover.

146

FIGURE 101. Rigging cover tied to footrests.

147

completes one half of the boat; the second half is constructed in the same manner.

(6) Both halves are placed in the water, side by side, and fastened together by looping a rope around their adjacent footrests.

(7) A lifting bar, or a strong pole, is placed on the saddles under the folded canvas at each end of the boat. The center bar is lashed securely to the center footrests with one end of a lair rope; the rope now is run to one end of the bar where it is used to secure the bar to the footrest; the rope is passed under the boat to the opposite end of the bar which again is secured to the footrest (fig. 102).

(8) The excess canvas at each end of the boat now is folded back over the saddle pads to provide protection. The boat is complete (fig. 103). With practice, it should not take over 15 minutes to construct such a boat which will support up to 2,500 pounds. With a load of 2,000 pounds, the freeboard will be 6 inches; with a load of 1,500 pounds, 12 inches. The foregoing data apply only to boats made with Phillips cargo saddles.

62. SELECTION OF BIVOUAC AREAS.

a. Bivouac areas should be selected with care and foresight so as to obtain the maximum facilities for the comfort of both men and animals. The requirements for such areas are:

(1) Concealment from air and ground observation.

(2) Cover from enemy fire and the elements.

(3) An adequate water supply.

(4) Firm footing. *River bottoms which may flood and steep hillsides which provide poor standing for the animals should not be selected.*

(5) Good drainage.

(6) Grazing.

(7) Areas large enough to provide adequate space for dispersion and free from such things as briars and poisonous plants.

FIGURE 102. Completed halves of boat lashed together.

b. When a stream cuts the route of march, and it is desired to bivouac nearby, the crossing should be made prior to establishing the bivouac.

63. BIVOUAC FORMATION. The arrangement of a bivouac must reduce to a minimum the danger from fire. The ground around kitchens and forges must be cleared of inflammable material, grass, and underbrush. Kitchens should be located so that dust from picket lines will not contaminate the food.

64. PICKET LINES FOR FIELD USE.

a. Picket lines for pack units are of two types: the ground picket line, and the raised picket line formed with lash ropes stretched between trees. Both are satisfactory and should be used in training so that men and animals will be accustomed to each. Tying first on one side of the line and then on the other, one animal may be tied every 24 inches.

b. Picket line areas should be level, free from rocks, stumps, undergrowth, and vines, and have natural drainage. If it is necessary to place the picket line on a hillside, the line should run up and down the slope so that the animals will have a level place on which to stand.

c. The ground picket line is stretched tight, flush with the ground, and is held in place by two steel pins, one at each end. If the ground is soft, it may be necessary to construct a deadman, Spanish windlass, or similar expedient to keep the stakes from pulling out. The picket line is tightened by means of a *pullback* described in e below. Animals should be tied to this line short enough to prevent them from getting their hind legs over the halter shanks, and long enough to allow them to stand with their heads in a natural position. Animals soon will accustom themselves to this line and stay clear of it.

d. The raised picket line is formed by stretching lash ropes (4 or 5 feet above the ground) between trees. It is tightened with a pullback described in e below. Animals should be tied with enough slack in the halter shank to allow them to eat hay off the ground or to lie down, but short enough to prevent them from getting their legs entangled in the rope.

e. The pullback used in tightening the picket line is formed as follows:

(1) Form a sheepshank knot in the picket line about 8 feet from the object to which the running end of the line is to be secured.

(2) Pass the running end of the line around the object to which it is to be secured and then back through the near eye of the sheepshank knot.

(3) Pull the running end back in the direction of the tree. This provides a two-to-one mechanical advantage, thus allowing the picket line to be stretched very tightly. (Picket lines should be kept taut at all times.) In forming the pullback for the ground picket line, a lash rope is secured to one end of the line and the pullback formed in it.

(4) The line now is secured by passing the running end around the object again and around all of the ropes of the pullback. It is tied by forming a slippery hitch directly behind the pullback.

f. A picket line guard should be provided for each section line. It is his duty to see that the animals are tied properly, prevent them from entangling and injuring themselves, keep the hay up under the center of the line and, in the absence of other personnel, remove feed bags when the animals have finished eating.

g. Picket lines should be policed daily while the unit is in bivouac. Manure must be disposed of by dispersion (spreading), burning, or composting, depending on how long the unit is to be bivouacked in one location. It is best to move the picket line every few days to allow the old standings to dry.

152

CHAPTER 10
EMERGENCY METHODS OF PACK TRANSPORT

65. PACKING CARGO ON RIDING SADDLES.

a. Riding saddles of most types may be used as emergency pack saddles. They generally are adaptable only to cargo loads.

b. The McClellan saddle, packer's saddle, or any stock type saddle may be packed readily by using a 30-foot rope to form the basket sling. The center of the rope is secured to the front arch or horn of the saddle and arranged as shown in figure 104. The loads are

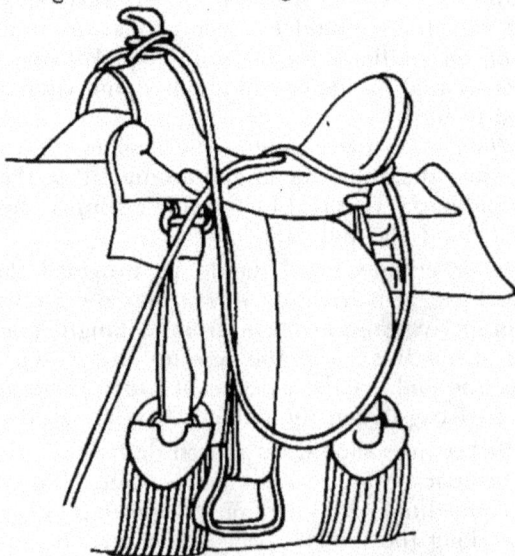

FIGURE 104. Basket sling rope on packer's or stock saddle. Initial rope formation.

placed on the saddle in either a vertical or horizontal position. They are secured as in the basket hitch (par. 41), except that no quarter ropes are drawn from under the standing ropes. The running end is tied to the standing rope with a sliding clove hitch (fig. 105).

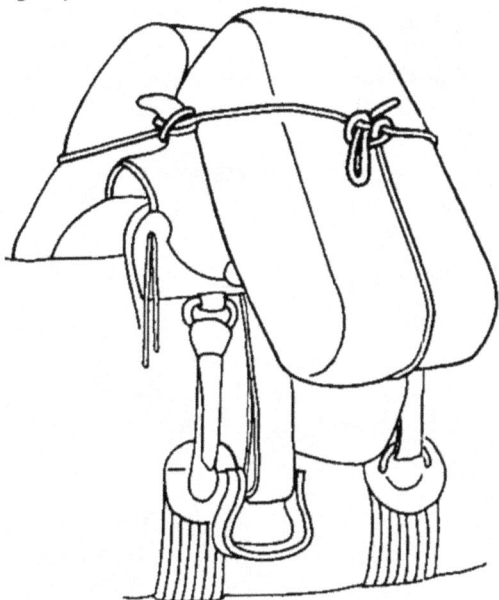

FIGURE 105. Basket sling, complete.

c. On the flat, or English type saddle, cargo may be packed by using the Wyoming diamond hitch as described in paragraph 66.

66. WYOMING DIAMOND HITCH.

a. This hitch is adaptable to any type of saddle, but is especially useful in packing on saddles which do not have footrests. A lash-cinch is used (fig. 106).

b. The hitch is formed as follows:

(1) The load being held in place by one of the normal slings, the near packer gets the lash-cinch, uncoils it

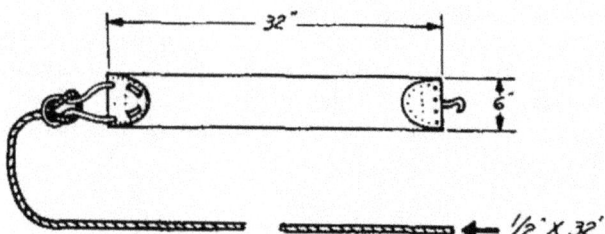

FIGURE 106. Lash-cinch (Wyoming diamond hitch).

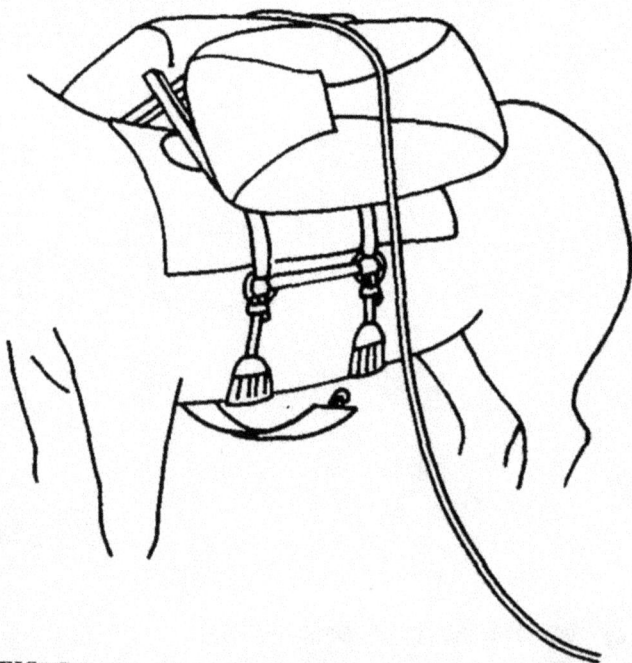

FIGURE 107. Cinch passed under animal's belly (Wyoming
diamond hitch).

and, holding the cinch, drops the remainder of the rope
to the ground on his right. He then passes the cinch
over the top of the cargo to the off packer who, in turn,
passes it back under the animal's belly (fig. 107).

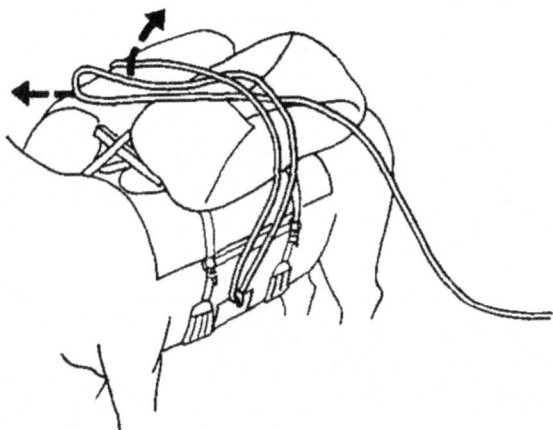

FIGURE 108. Loop passed beneath standing rope (Wyoming diamond hitch).

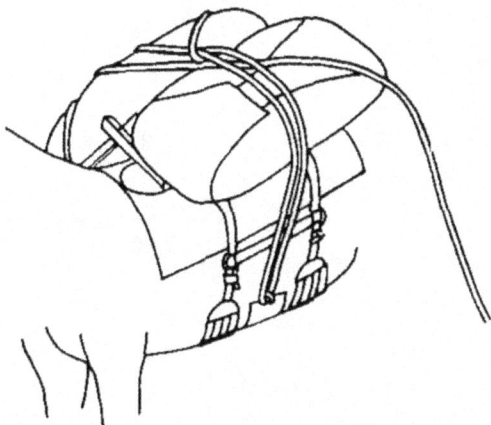

FIGURE 109. Loop spread by off packer (Wyoming diamond hitch).

(2) The near packer passes the running rope through the hitch hook, front to rear, and pulls the lash-cinch snug around the load and animal. He then forms a loop in the running end of the rope, and passes it, rear

to front, beneath the standing rope at a point near the top of his side load (fig. 108). The off packer enlarges the loop and spreads it, front and rear, so that it forms end ropes to pass around the ends of his side load (fig. 109).

(3) The near packer now reaches between the standing ropes on the near load, grasps the running rope and taking slack from the free end, forms a loop. He enlarges the loop and spreads it, front and rear, so that it forms end ropes to pass around both ends of his side load (fig. 110).

(4) The near packer now tightens ("hits") the rear standing rope while the off packer takes slack. Retaining all slack, the off packer guides the rope of his loop

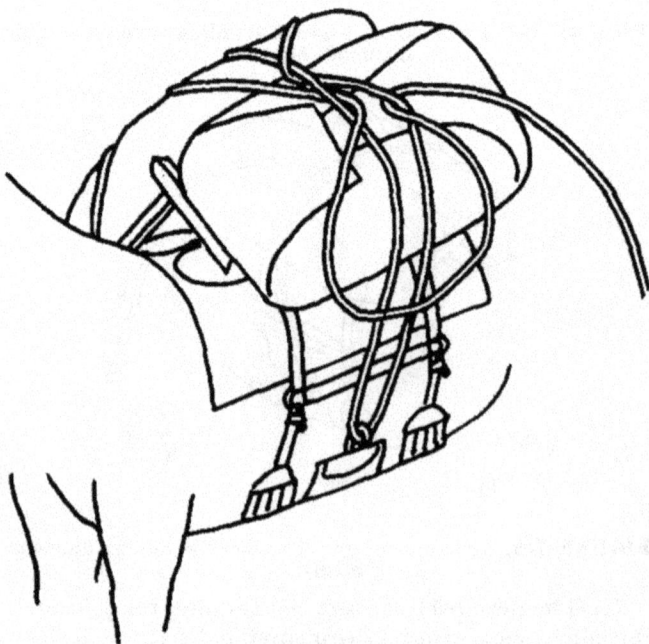

FIGURE 110. Loop pulled out by near packer (Wyoming diamond hitch).

around the rear corner of his side load, along the bottom, and up around the front corner. To remove all slack, the near packer now pulls hard to the front on his front end rope. Retaining all slack, he guides the rope of his loop around the front corner of his side load, along the bottom, and up around the rear corner.

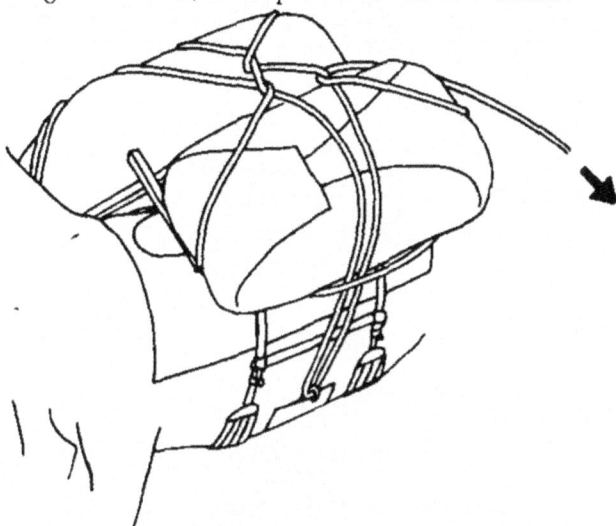

FIGURE 111. Wyoming diamond hitch formed and ready to be tied.

(5) The near packer now pulls hard on the running rope, removing all slack (fig. 111). He then secures it, between the loads, to the rear standing rope with a slippery hitch. All loose ends are tucked away.

67. IMPROVISED PACK SADDLES.

a. The sawbuck or cross-tree saddle consists of two saddle boards or side bars connected at the front and rear by crosspieces shaped like the letter "X." It should be equipped with a breast strap, breeching, quarter straps, cinches and latigos, and a saddle pad or blanket. Because of its simplicity, it may be constructed by unit carpenters and saddlers.

158

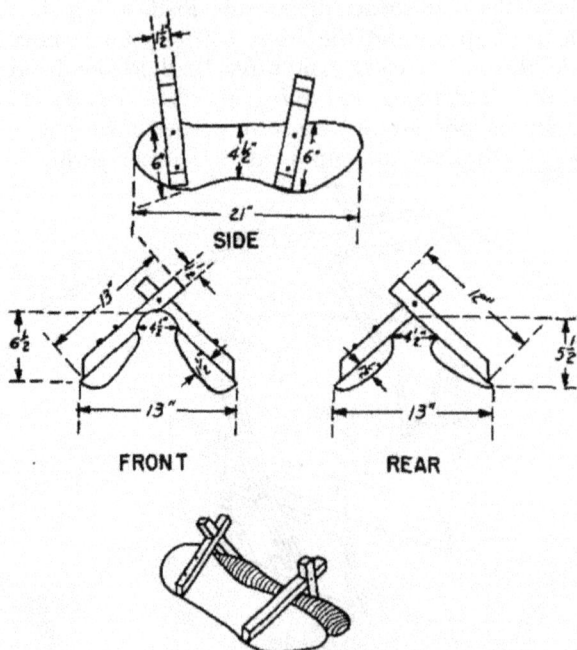

FIGURE 112. Sawbuck saddle tree.

b. A similar saddle may be constructed by substituting horseshoes for the crosspieces of the sawbuck (fig. 114).

c. When improvising saddles, care must be exercised to insure that the side bars will bear with equal pressure on the weight-carrying surfaces of the animal's back. The side bars should be made of seasoned soft, light, fine-grained wood, such as white pine or birch. The crosspieces of the sawbuck should be made of well-seasoned hardwood, such as oak, hickory, or elm.

68. CONSTRUCTION OF TRAVOIS.

a. A travois is a primitive vehicle for transporting loads. It consists of two trailing poles which serve as shafts for the animal and a platform or net for the load.

FIGURE 113. Suggested rigging for sawbuck saddle.

HORSESHOES

GENERAL DIMENSIONS
SAME AS SAWBUCK

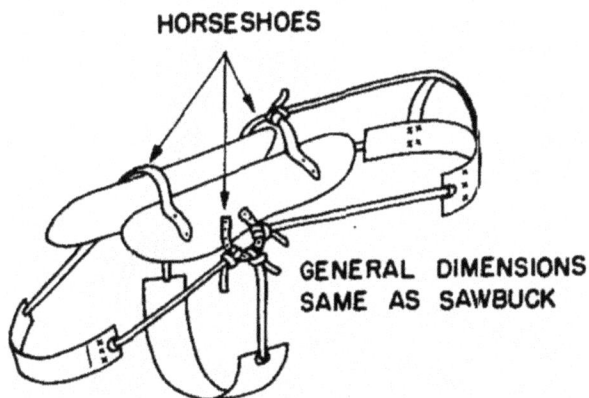

FIGURE 114. Pack saddle improvised with horseshoes as
crosspieces.

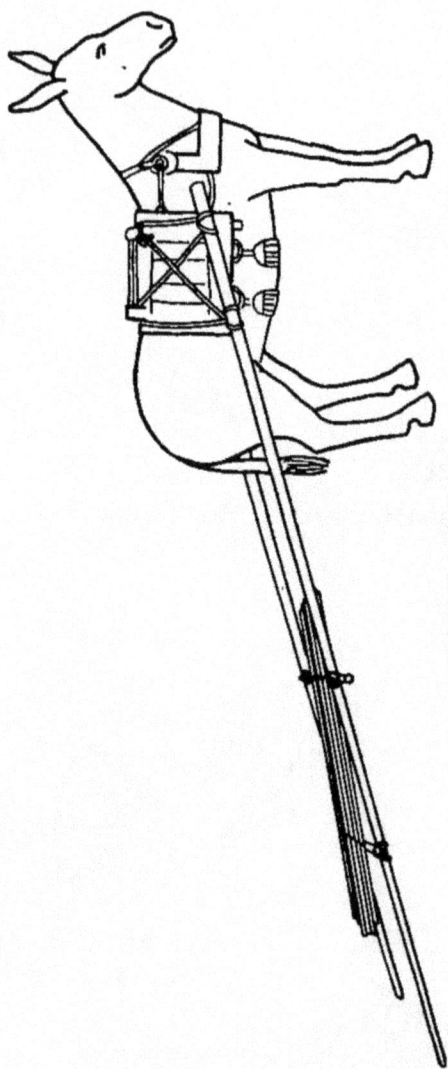

FIGURE 115. Travois.

b. The poles of the travois should be straight, strong, light, and from 14 to 20 feet long, depending upon the size, weight, and type of load to be transported. The pole's diameter at the smaller end should be not less than 2 inches.

c. The large ends of the poles are secured to the near and off sides of the animal by lashing them to a saddle, or an improvised pulling harness (fig. 115).

d. The platform, normally from 2 to 3 feet wide, may be made of canvas, boards, or small logs. Its front edge should be approximately 3 feet in rear of the animal's hind legs.

e. Spacers placed between the poles at the front and rear of the platform provide rigidity.

f. A travois team should consist of three packers, one packer to lead the animal and two to follow and assist in crossing ditches, streams, and other obstacles.

INDEX